Gabriele Rizzi

Are we living in another Migration Period?

Pandemics, migrations and the environmental link. A focus on Europe

Preface
Mario Giro
Former Vice-minister of Foreign affairs
and Professor of international relations

«L'ERMA» di BRETSCHNEIDER
Roma – Bristol

Gabriele Rizzi

Are we living in another Migration Period?

Via Marianna Dionigi, 57 70 Enterprise Drive, Suite 2
00193 Roma – Italia Bristol, CT 06010 – USA
www.lerma.it lerma@isdistribution.com

Sistemi di garanzia della qualità
UNI EN ISO 9001:2015

Sistemi di gestione ambientale
ISO 14001:2015

Gabriele Rizzi - *Are we living in another Migration Period?* : Roma: «L'ERMA» di BRETSCHNEIDER®, 2022 - 204 p.; 24 cm

ISBN Paperback: 978-88-913-2511-2
ISBN Pdf: 978-88-913-2513-6

DOI: 10.48255/M.G.Rizzi.9788891325136

Stampato nel rispetto dell'ambiente su carta proveniente da zone a deforestazione controllata

Contents

Prefazione

Uomini e popoli si sono sempre spostati nel corso della storia. Quello che sul breve periodo può sembrare un evento acuto e talvolta disastroso, se guardato sul lungo periodo risulta essere un fenomeno ordinario e ripetibile di mobilità umana. In questo originale e approfondito saggio Gabriele Rizzi connette tale esperienza umana con l'ambiente circostante, cioè con gli aspetti antropologici e ambientali in cui tale mobilità di sviluppa, incluse le pandemie. Lo fa mediante la comparazione di epoche diverse (le invasioni barbariche – o il periodo della grandi migrazioni – e il tempo attuale) che assumono un volto familiare di somiglianza tra loro ma anche differenze nette, in particolare a riguardo dei cambiamenti ambientali (ma non pandemiche a causa della peste). L'approccio comparativo utilizzato dall'autore permette di far risaltare similitudini e diversità, essenziali nel costruire un modello predittivo sul futuro. La domanda del saggio è se ci troviamo davanti ad un altro periodo di grandi migrazioni o spostamenti di popolazioni come avvenne all'epoca della fine dell'Impero romano d'Occidente. Ciò significa analizzare, come fa l'autore, il tipo di cause e conseguenze che spingono le popolazioni a spostarsi. In effetti nell'attuale ciclo di globalizzazione (che non è l'unica nella storia) numerosi focolai di crisi politica, ambientale ed economica spingono le persone a trasferirsi in maniera imprevista. Ciò colpisce particolarmente alcune aree del pianeta. Non si tratta soltanto di fenomeni economici: in una prima fase la globalizzazione ha messo in crisi la democrazia liberale (mediante il divorzio mercato-democrazia); nella fase pandemica successiva anche la globalizzazione stessa è entrata in crisi come osserviamo attualmente. Analizzando lo scenario possiamo dunque analizzarne – come fa lo studio del Rizzi – le conseguenze antropologiche. In alcuni casi la dimensione degli eventi è tragica: in Siria tra IDP e rifugiati è stata colpita metà della popolazione che ha dovuto lasciare le proprie case. Si tratta di un enorme push factor che ha aperto un flusso anche per altri e non solo in Medio Oriente. I rifugiati siriani hanno "votato con i piedi" contro la loro guerra ma anche contro la comunità internazionale che non ha saputo o voluto fermarla. Ci dobbiamo aspettare lo stesso dai libici se non si arriva presto alla pace? Molti puntano all'Europa o agli Usa: infatti malgrado i pregiudizi e la differenza culturale ciò che tutti percepiscono è che in Europa e negli Stati Uniti si vive meglio, la vita è più rispettata. Ma le società occidentali difficilmente ammettono che il rispetto della vita e dei diritti possa divenire un pull factor. Le crisi nel Mediterraneo orientale ma anche del Medio Oriente, del Golfo, del Caucaso ecc. sono crisi di convivenza: la coabitazione si infrange con violenza contro lo tsunami delle identità etnico-religiose e contro gli interessi politici. In un universo in cui le frontiere tornano ad essere una posta in gioco, i migranti soffrono ma trovano sempre nuove strade. Nel disordine globale si aprono vie innovative. Quando ciò avviene nel disordine e senza la regolazione degli Stati, ecco apparire i trafficanti che organizzano il commercio di "carne umana". Poi c'è lo scandalo dell'isolamento: si pensi a Lesbos, alla Bosnia o alla frontiera polacco-bielorussa: persone senza diritti nella terra dei diritti, bloccate e senza scampo. A tale caos si aggiunge la pandemia che non cessa di colpire e pare non voler terminare tanto presto. Tanti fenomeni che causano altrettante ondate migratorie che a loro volta provocano paure e idiosincrasie con la nascita di nuovi focolai di crisi. Negli anni si è creato un circolo vizioso. Per l'Europa, più esposta ai flussi provenienti dal Nord Africa, dai Balcani o dal Mediterraneo, l'unica soluzione consiste nel rendere multilaterale la questione dei flussi. Come cita il saggio di Rizzi, attualmente l'idea più innovativa, messa in opera in Italia dal 2016, sono i corridoi umanitari promossi dalla Comunità di Sant'Egidio assieme alla Federazione Evangelica italiana, alla Tavola Valdese e, in un secondo momento, alla Caritas italiana. La sicurezza sia di chi è accolto che di chi accoglie mostra l'impatto *win-win* del modello. Una volta giunti in Italia i profughi sono accolti dai promotori e, in collaborazione con altri partner, vengono ospitati e integrati dalla società stessa. C'è a questo proposito una grande mobilitazione di chi si occupa di cercare le sistemazioni, l'integrazione scolastica, lavorativa e così via. Viene offerta un'inclusione nel tessuto sociale e culturale attraverso l'apprendimento della lingua, la scolarizzazione ed altre iniziative. Il modello è quello dell'accoglienza diffusa, personalizzata, secondo un percorso "adozionale", che coinvolge le collettività locali. Sinora nell'ambito del programma l'offerta di accoglienza supera la domanda: un segnale chiaro e in controtendenza rispetto agli umori negativi che spesso si creano nelle società europee. Il modello dei corridoi è stato adottato anche in Francia e in Belgio ed rappresenta l'unica via legale attualmente operativa in Europa. La sfida che propone questo studio è quella di provare a comprendere lo spessore antropologico e umano dello scenario complesso degli spostamenti delle popolazioni, ampliando lo scenario della visione futura. Concludo a questo proposito citando le parole di un prete messicano che si batte per il rispetto della vita sulla frontiera del grande flusso che dal Centroamerica attraversa il suo paese per giungere negli Stati Uniti. Afferma Solalinde che *"i migranti sono un segno dei tempi. Sono vittime del neoliberismo selvaggio che ha divorato il loro paese d'origine e li ha costretti a lasciarlo. In questo senso, sono testimoni di un mondo in disfacimento, ne portano le ferite nella loro carne. Al contempo, però, i migranti sono i pionieri del futuro. Anticipano, con la loro ostinata resistenza, la possibilità di una nuova società. Perché? Perché hanno il coraggio di rischiare. [...] Il loro viaggiare, invincibile e dolente, rammenta a noi, ormai accomodati e aggrappati alle nostre certezze, che siamo tutti pellegrini. Tutti siamo migranti* ". ("I Narcos mi vogliono morto", A. Solalinde e L. Capuzzi, 2017, EMI.).

Mario Giro

già vice ministro degli Esteri
e Professore di Relazioni internazionali

Preface

Men and peoples have always moved throughout history. What in the short term may seem an acute and sometimes disastrous event, when looked at in the long term turns out to be an ordinary and repeatable phenomenon of human mobility. In this original and in-depth essay, Gabriele Rizzi connects this human experience with the surrounding environment, that is, with the anthropological and environmental aspects in which such mobility develops, including pandemics. He does so by comparing different epochs (the barbarian invasions – or the Migration period – and the present time) that take on a familiar face of similarity but also clear differences, particularly with regard to environmental changes (but not pandemics due to the plague). The comparative approach used by the author makes it possible to bring out similarities and differences, which are essential in constructing a predictive model of the future. The question of the essay is whether we are facing another period of great migration or displacement of populations as occurred at the time of the end of the Western Roman Empire. This means analysing, as the author does, the kind of causes and consequences that drive populations to move. Indeed, in the current cycle of globalisation (which is not unique in history), numerous political, environmental and economic crises are driving people to move unexpectedly. This particularly affects certain areas of the planet. It is not only economic phenomena: in the first phase, globalisation undermined liberal democracy (through the market-democracy divorce); in the next pandemic phase, globalisation itself also went into crisis, as we are currently seeing. By analysing the scenario we can then analyse – as Rizzi's study does – its anthropological consequences. In some cases the scale of events is tragic: in Syria, between IDPs and refugees, half the population has been affected and has had to leave their homes. This is a huge push factor that has opened up a flow for others and not only in the Middle East. The Syrian refugees "voted with their feet" against their war but also against the international community that was unable or unwilling to stop it. Should we expect the same from the Libyans if peace is not achieved soon? Many point to Europe or the US: in fact, despite prejudices and cultural differences, what everyone perceives is that life is better in Europe and the US, life is more respected. But Western societies find it difficult to accept that respect for life and rights can become a pull factor. The crises in the Eastern Mediterranean, but also in the Middle East, the Gulf, the Caucasus, etc., are crises of coexistence:

cohabitation is violently shattered by the tsunami of ethnic-religious identities and political interests. In a world where borders are once again at stake, migrants suffer but always find new paths. In the global disorder, innovative ways open up. When this happens in disorder and without the regulation of states, traffickers appear, organising the trade in 'human flesh'. Then there is the scandal of isolation: think of Lesbos, Bosnia or the Polish-Belarusian border: people without rights in the land of rights, stranded and with no way out. In addition to this chaos, there is the pandemic which is still striking and does not seem to want to end any time soon. So many phenomena that cause as many waves of migration, which in turn provoke fears and idiosyncrasies with the birth of new hotbeds of crisis. Over the years, a vicious circle has been created. For Europe, which is more exposed to flows from North Africa, the Balkans or the Mediterranean, the only solution is to make the issue of flows multilateral. As Rizzi's essay mentions, currently the most innovative idea, put in place in Italy since 2016, are the humanitarian corridors promoted by the Community of Sant'Egidio together with the Italian Evangelical Federation, the Tavola Valdese and, at a later stage, the Italian Caritas. The safety of both those being welcomed and those receiving shows the *win-win* impact of the model. Once in Italy, refugees are welcomed by the promoters and, in cooperation with other partners, are hosted and integrated by society itself. There is great mobilisation of those involved in finding accommodation, school and work integration and so on. Inclusion in the social and cultural fabric is offered through language learning, schooling and other initiatives. The model is that of widespread, personalised reception, according to an "adoption" process involving the local communities.

To date, the supply of reception facilities has exceeded demand under the programme, which is a clear signal that runs counter to the negative moods that are often created in European societies. The corridor model has also been adopted in France and Belgium and is the only legal route currently operating in Europe. The challenge posed by this study is to try to understand the anthropological and human depth of the complex scenario of population movements, broadening the scenario of the future vision. I would like to conclude by quoting the words of a Mexican priest who fights for respect for life on the frontier of the great flow that crosses his country from Central America to the United States. Solalinde says that 'migrants are a sign of the times. They are victims of the savage neo-liberalism that has devoured their country of origin and forced them to leave it. In this sense, they are witnesses of a world in decay, they carry its wounds in their flesh. At the same time, however, migrants are the pioneers of the future. They anticipate, with their stubborn resistance, the possibility of a new society. Why? Because they have the courage to take risks. [...] Their journey, invincible and painful, reminds us, now settled and clinging to our certainties, that we are all pilgrims. We are all migrants". ("The Narcos want me dead", A. Solalinde and L. Capuzzi, 2017, EMI.).

Mario Giro
Former Vice-minister of Foreign affairs
and Professor of international relations

Introduction

Pandemics, migrations, and climate change: a summary of 2020, one could say. However, we are not talking about our historical period. The same events, in fact, occurred during the so-called *Migration Period*, a historical period from the 3rd to the 9th century, during which populations across Eurasia left their original homeland. This was due to several factors, but more and more studies are revealing that climate change played an important role. Importantly, the same is happening now. It is sometimes said that climate change will cause people to leave their homes in the future. However, evidence reveals that changing climate conditions have already played an important role in a major conflict that has been at the root of a large migration flow: the Syrian civil war. Moreover, zoonotic spillovers, responsible for Covid-19 and other twenty-first century pandemics, are also fostered by habitat disruption, one of the main drivers of climate change, along with CO2 emissions.

Therefore, are we on the threshold of another Migration period, an era of great migratory flows, pandemics and climate change? The coincidence of the Covid-19 outbreak and the migration crisis of recent years seems to suggest so. We believe, in fact, that climate change will become the main driver of migrations and conflict in the future, while nowadays it is already a "threat multiplier", being at the basis of droughts, floods and rising tensions in many countries of the Middle East and Africa. Europe is therefore already witnessing the arrival of the first "climigrants", but it could be argued that in the future more and more people will decide to leave their countries due to the effects of changing weather conditions and even Europeans may decide to leave for countries with more favourable conditions.

The current study addresses the issue through a comparison between our era and the Migration Period. Such a comparative approach is useful to grasp common trends and similarities between different periods and at the same time it can help to predict possible evolutions. In the course of the analysis we will mainly focus on the following questions: how do people react to pandemics? Were the effects of the Justinian plague different from the ones of the Covid-19? And what are the reasons for people to leave their country? So, can policymakers learn from the past and can we suggest possible policies?

Remarkably, we decided to focus on the Migration period because it is also known in Southern European countries as the period of the "barbarian invasions". Such a marked difference (barbarian invasions/migrations) is extremely interesting because of the different significance the same period can have in different cultures or countries. We therefore decided to focus on early medieval Europe to study the migration waves of that period with the aim of approaching the present ones as historical processes triggered by important causes that we have to deal with. Moreover, the presence of pandemics in both periods gives the possibility to compare the two periods, focusing also on the interactions between pandemics, migration, climate change and societal response.

The analysis will also be based on two key points. First, the study will address the concept of globalisation: in our view, globalisation is intrinsic to humanity, although it can certainly vary between societies and historical periods. Indeed, there is no doubt that our society is more globalised than that of the migration period. However, in the past, globalization travelled along trade and migration routes, and so did Justinian's plague. The same can be said, for example, for Covid-19. However, as our world is much more connected than the early medieval one, trends and processes are correspondingly accelerated. For example, while the Justinian plague took a year to reach Constantinople – the most important Mediterranean city at the time – from Pelusium (where the first outbreak was recorded), Covid-19 probably reached New York from Wuhan in less than a day. This gives an idea of the importance of analysing and comparing historical trends in order to predict possible immediate consequences.

The second major issue will be collaboration between sciences: in such a globalised world, a holistic approach is needed, especially among policymakers. In the current study, we will be looking at medicine, epidemiology, physics, sociology, economics, geopolitics and, of course, history. This approach will therefore highlight the need to adopt a point of view that embraces the different branches of science; both in the past and in our time, this is exemplified, for example, by the fact that a small bacterium or virus can have major geopolitical effects.

The analysis will be divided into three chapters. In the first chapter we will focus on the period of migrations, analysing the effects of the Justinian plague on society, its possible spread, and the mystery of its origin. We will then move on to the migratory movements that characterised the historical period, describing the migratory routes and the causes of migration. Finally, we will deal with the climate change that took place at that time, arguing possible connections between it and the migrations of Germans, Slavs and Avars, as well as with Justinian's pandemic.

The second chapter will deal with the contemporary era: after introducing the pandemics of the 21st century, we will focus on the effects of Covid-19 on contemporary society, from an economic and geopolitical point of view. The following paragraph, on the other hand, is dedicated to migration routes, but also returns to the issue of the current pandemic, showing how people in developing countries are losing their jobs, and therefore may decide to migrate in the near future. Furthermore, in the fifth paragraph, we argue that the current climate change is already responsible for migration and that the past and current pandemics are also triggered by the destruction of natural habitats.

In the final chapter we finally address the research question: are we experiencing another period of migration? The chapter mainly compares contemporary and early medieval Europe, focusing on similarities and common trends. We then investigate possible evolutions and challenges, introduce the international climate agreements and the mitigation strategies to address climate change, as well as the European Green Deal. Finally, we suggest possible policies to address climate change and related issues in the short, medium and long term.

I. The Migration Period

1.1 The 'Justinian plague' and the first plague pandemic

What became known as the 'Justinian plague' first appeared in CE 541 in the Mediterranean port of *Pelusium,* located in the easternmost bank of the Nile. Its outbreak is documented by several contemporary sources which have no doubt in placing the – historical – beginning of the "first plague pandemic" in Egypt. The certainties, however, end here. For instance, scientists proved that *Yersinia pestis* was the bacterium responsible of the pandemic only in 2013[1] and a huge debate regarding the real number of the losses is still going on, as it will be illustrated below. It is not yet known why the pandemic ended suddenly, after several outbreaks over 200 years[2]. Moreover, we still do not know the exact route which led the plague to the shores of the Mediterranean Sea, and this is bound to remain a mystery unless new discoveries provide new certainty[3].

The best approach therefore seems to start our analysis from the historical sources. Among the most important contemporary ones we include Procopius (c. 500 – c. 570), author of the *History of the Wars,* John of Ephesus (c. 507 – c. 588) author of the *Ecclesiastical History,* Evagrius Scholasticus (c. 536 – c.594), author of the *Ecclesiastical History* and Gregory of Tours, (c. 538 – 594), author of the *History of the Franks*. We will also mention Paul the Deacon (c.720 – 796/9), author of the *History of the Lombards*.

Justinian, the emperor from which the plague took the name, was in his fourteenth year of reign and in that year, he also had to deal with the invasion of the Ostrogoths in Italy and the Persians pushing in modern day Georgia. In fact, Europe was in the midst of the Migration Period and was crossed by the armies of the Germanic and Asian tribes; moreover, four years before, an obscure event had covered the skies of the Continent: from then on, 536/7 will be known as the "year without summer[4]". We will deal with both issues later. Nonetheless, this gives the idea of an entire period which has been interpreted as a *caesura* between the Roman era and the Middle Ages[5].

---

[1] Harbeck M, Seifert L, Hänsch S, Wagner DM, Birdsell D, Parise KL, et al, *Yersinia pestis DNA from Skeletal Remains from the 6th Century AD Reveals Insights into Justinianic Plague*, PLoS Pathog 9, 5, 2013

[2] McCormick, *Toward a Molecular History of the Justinianic Pandemic,* in Plague and the end of antiquity, the pandemic of 541-750, edited by Lester K. Little, Cambridge University Press in association with The American Academy in Rome, 2007.

[3] Sallares R., *Ecology, Evolution, and Epidemiology of Plague* in Plague and the end of antiquity, the pandemic of 541-750, cit.

[4] Gunn, J.D., *The years without summer: tracing AD 536 and its aftermath,* British Archaeological Reports International, Archaeopress, 2000.

[5] Meir, M., *The 'Justinianic Plague': the economic consequences of the pandemic in the eastern Roman empire and its cultural and religious effects*, Early Medieval Europe, 24,3, pp. 267-292, 2016.

Considering that, as said before, all the contemporary sources agree on the fact that the first outbreak was recorded in Egypt[6] or in the Red Sea[7], it is interesting to underline the fact that they also agree that it spread via sea, that is to say thanks to the main long-distance mean of transport of the time. This is underlined by both ancient sources like Gregory of Tours ("a ship from Spain [...] unhappily brought the seed of this disease"[8]) and Procopius ("[...] this disease always took its start from the coast, and from there went up to the interior") and modern research[9].

Alexandria, the main port of the Mediterranean at the time, was hit the same year of Pelusium. Then, the plague moved from Egypt to the rest of the known world, reaching as far as Ireland in 544; in 542 it arrived in Palestine, North Africa, Sicily and Anatolia, in 543 it arrived in Rome, Spain and Northern France and the following year it finally reached the North Sea[10].

In Constantinople, the capital of the Eastern Roman Empire, the plague arrived in 542. Among the others, at the court of the Emperor Justinian there was the historian Procopius. He therefore witnessed the effects and the symptoms of the plague and gave a precise description of them:

> "And in the second year [the disease] reached Byzantium in the middle of spring, where it happened that I was staying at that time [...] But with the majority it came about that they were seized by the disease without becoming aware of what was coming either through a waking vision or a dream. And they were taken in the following manner. They had a sudden fever, some when just roused from sleep, others while walking about, and others while otherwise engaged, without any regard to what they were doing. And the body shewed no change from its previous colour, nor was it hot as might be expected when attacked by a fever, nor indeed did any inflammation set in, but the fever was of such a languid sort from its commencement and up till evening that neither to the sick themselves nor to a physician who touched them would it afford any suspicion of danger. It was natural, therefore, that not one of those who had contracted the disease expected to die from it. But on the same day in some cases, in others on the following day, and in the rest not many days later, a bubonic swelling developed; and this took place not only in the particular part of the body which is called "boubon" that is, below the abdomen, but also inside the armpit, and in some cases also beside the ears, and at different points on the thighs[11]"

[6] "It started among the Egyptians" (Procopius, *History of the Wars, II, 22*).

[7] "It took its rise from Aethiopia, as is now reported, and made a circuit of the whole world in succession" (Evagrius, *Ecclesiastic History,* XXIX).

[8] Gregory of Tours, *History of the Franks*, translation of Earnest Brehaut. It is also interesting to observe that other authors such as John of Ephesus and Procopius claim that people were having visions of bronze boats roaming over the sea with a crew of headless men ("When this plague was passing from one land to another, many people saw shapes of bronze boats and (figures) sitting in them resembling people with their heads cut off. Holding staves, also of bronze, they moved along on the sea and could be seen going whith ersoever they headed. These figures were seen everywhere in a frightening fashion, especially at night. Like flashing bronze and like fire did they appear, black people without heads sitting in a glistening boat and traveling swiftly on the sea, so that this sight almost caused the souls of the people who saw it to expire").

[9] Little, K., *Life and Afterlife of the First Plague Pandemic in Plague and the end of Antiquity*, in Plague and the end of antiquity, the pandemic of 541-750, cit.

[10] Keller, M., Spyrou, M.A., et al., *Ancient Yersinia pestis genomes from across Western Europe reveal early diversification during the First Pandemic (541–750)*, PNAS, 116, 25, 12363-12372, 2019.

[11] Procopius, *History of the wars*, Book II. 22, translated by H. B. Dewing, Loeb Classical Library 48. Cambridge, MA: Harvard University Press, Cambridge, 1914.

Similar symptoms are also described by the other important source of the time, Evagrius, who even fell ill and survived:

"The plague was a complication of diseases: for, in some cases, commencing in the head, and rendering the eyes bloody and the face swollen, it descended into the throat, and then destroyed the patient. In others, there was a flux of the bowels: in others, buboes were formed, followed by violent fever; and the sufferers died at the end of two or three days, equally in possession, with the healthy, of their mental and bodily powers. Others died in a state of delirium, and some by the breaking out of carbuncles. Cases occurred where persons, who had been attacked once and twice and had recovered, died by a subsequent seizure"[12].

We must consider that the Greek and Latin sources (Procopius and Evagrius) were both aware of previous epidemics[13] and that they had also probably read Thucydides' accounts of the Athens' pestis[14]. This explains why they tried to give a rational explanation and why they were interested in describing the works of the physicians, who tried to investigate the nature of the disease, although not understanding much, according to Procopius:

"Now some of the physicians who were at a loss because the symptoms were not understood, supposing that the disease centred in the bubonic swellings, decided to investigate the bodies of the dead. And upon opening some of the swellings, they found a strange sort of carbuncle that had grown inside them [...]. I am able to declare this, that the most illustrious physicians predicted that many would die, who unexpectedly escaped entirely from suffering shortly afterwards, and that they declared that many would be saved, who were destined to be carried off almost immediately. So it was that in this disease there was no cause which came within the province of human reasoning; for in all cases the issue tended to be something unaccountable"[15]

From the accounts of the time, we can see that there was speculation about how the disease could spread:

"For some perished by merely living with the infected, others by only touching them, others by having entered their chamber, others by frequenting public places. Some, having fled from the infected cities, escaped themselves, but imparted the disease to the healthy. Some were altogether free from contagion, though they had associated with many who were afflicted, and had touched many not only in their sickness but also when dead.[16]"

---

[12] Evagrius, *Ecclesiastic History, History Of The Church, In Six Books, From A. D. 431 To A.D. 594, A New Translation From The Greek, with an account of the author and his writings,* Samuel Bagster And Sons, London, 1846 Chapter XXX.

[13] K. Little, *Life and Afterlife of the First Plague Pandemic in Plague and the end of Antiquity*, edited by Lester K. Little. Cambridge: Cambridge University Press, in association with the American Academy in Rome, 2006.

[14] Maddicot, J.R., *Plague in seventh-century England*, Past & Present, in Plague and the end of antiquity, the pandemic of 541-750, cit; "in some respects similar to that described by Thucydides", Evagrius, *Ecclesiastic History*, chapter XXIX,cit.

[15] Procopius, *History of the Wars* 7 Vols., translated by H. B. Dewing, cit.

[16] Evagrius, *Eccleasiastic History,* Chapter XXIX, cit.

What struck the historians of the time was the fact that while the disease seemed to have spread all over the world[17] and to infect through contact, they could not understand why several houses or areas were left uninfected and why some people were not contaminated:

> "In some instances, having infected a part of a city, it left the remainder untouched; and frequently in an uninfected city one might remark a few households excessively wasted; and in several places, while one or two households utterly perished, the rest of the city remained unvisited: but, as we have learned from careful observation, the uninfected households alone suffered the succeeding year [...] For neither physicians nor other persons were found to contract this malady through contact with the sick or with the dead, for many who were constantly engaged either in burying or in attending those in no way connected with them held out in the performance of this service beyond all expectation, while with many others the disease came on without warning and they died straightway."[18]

Furthermore, Procopius argues that the cities that were left untouched "were seized with disorder", a description of mass hysteria that is also described by other sources, which point out that those who remained sane started to behave like dogs or sheep[19]. Therefore, the conclusion of Procopius is:

> "I am unable to say whether the cause of this diversity of symptoms was to be found in the difference in bodies."

Moreover, we must not forget that, at the time, many attributed the disease to the magical world: indeed, the historians wrote of visions of headless supernatural beings who transmitted the disease[20] or about bronze people rowing bronze boats[21]. On the other hand, physicians attributed the plague to "bad airs", but this explanation was also used for malaria[22].

We know now that the disease was caused by the bacterium *Yersinia Pestis*[23], which was also responsible for the "second and the third plague pandemic and is still with us, appearing from time to time in several locations known as *bacterial foci*, i.e. natural locations in which the bacterium is enzootic, native to animals[24]. Indeed, the disease affects several species of rodents and it is transmitted to humans by bites of infected

[17] "It embraced the entire world [...] For it left neither island nor cave nor mountain ridge which had human inhabitants" (Procopius).

[18] Evagrius, *Eccleasiastic History*, Chapter XXIX, cit.

[19] Atkinson, J., *The Plague of 542: not the birth of the Clinic*, Acta Classica, Vol. 45, pp. 1-18, 2002.

[20] Procopius, *History of the Wars*, cit.

[21] John of Ephesus, *Ecclesiastic History*, cit.

[22] Sallares R., *Ecology, Evolution, and Epidemiology of Plague* in Plague and the end of antiquity, the pandemic of 541-750, cit.

[23] Harbeck M, Seifert L, Hansch S, Wagner DM, Birdsell D, et al., *Yersinia pestis DNA from Skeletal Remains from the 6th Century AD Reveals Insights into Justinianic Plague*, cit.

[24] Eroshenko GA, Nosov NY, Krasnov YM, Oglodin YG, Kukleva LM, Guseva NP, et al., *Yersinia pes tis strains of ancient phylogenetic branch 0.ANT are widely spread in the highmountain plague foci of Kyrgyzstan*, PLoS ONE 12, 10, 2017.

fleas which leave the body of the animals[25]. This explains why several houses were left untouched: sanitary conditions must have varied between houses, or in certain cases people probably lived nearer to food supplies, i.e. closer to what attracts gnawers. This also explains why, according to the historical sources, poor people were affected more than others[26]: their living conditions at the time must have been appalling. Interestingly, the historian McCormick, author of *Tracking mass death during the fall of Rome* even argues that "butchers and slaughter houses attract rats, whereas some evidence suggests that the intermittent noise of carpenters and metalworkers drives them away"[27].

1.1.1 Mortality and demographic effects

Apart from the "poor", the plague does not seem to have differentiated by sex or even age; this is, at least, what we can gather from the historical sources[28], and is confirmed by several contemporary studies: according to McCormick, indeed, a study found that while "sex seems not to have strongly affected the risk of death from plague, another study argues that, unsurprisingly, previous frailty did increase that risk somewhat". Mortality does not appear to have varied in relation to where it hit, according to Procopius, who states that the plague victims corresponded "exactly to the number destroyed at the earlier time"; he also adds that the pestilence lasted four months in the city of Constantinople and:

> "at first the deaths were a little more than the normal, then the mortality rose still higher, and afterwards the tale of dead reached five thousand each day, and again it even came to ten thousand and still more than that[29]."

Furthermore, Procopius in his *Anecdota* or *Secret History* argues that the total amount of the deaths was about half the population of the city[30]. Nonetheless, there is no agreement on the issue: while several authors such as Kyle Harper suggest that the Justinian plague killed half the population of the Eastern Roman Empire during the plague waves with devastating effects in the society of the time[31], others such as historians Mordechai and Eisenberg rejects this interpretation[32]. As for the latter interpretation, the authors of *Rejecting Catastrophe: The case of the Justinian Plague,* claim

---

[25] Retief, F.P., *The epidemic of Justinian (AD 542): a prelude to the Middle Ages*, Acta Theologica, Supplementum 7, 2005.

[26] "poor, for they died first" Michael the Syrian in Rosen, W., Justinian's Flea, Plague, Empire, and the Birth of Europe, Viking Penguin, 2007.

[27] McCormick, M., *Tracking mass death during the fall of Rome's empire (I). Journal of Roman Archaeology*, 28, pp. 325-357, 2015.

[28] "However, they say that three women in confinement survived though their children perished" (Procopius *History of the Wars*).

[29] Procopius, *History of the Wars*, cit.

[30] "And afterwards came the plague as well, mentioned by me before, which carried off about one-half of the surviving population" (Procopius, *Secret History*, XVIII, 44, translation of Henry Bronson).

[31] Harper, K., *The Fate of Rome, Climate, Disease and the End of an Empire*, The Princeton History of the Ancient World, Princeton University Press, 2017.

[32] Mordechai, L., Eisenberg, M., *Rejecting Catastrophe: The Case of the Justinianic Plague*, Past & Present, Volume 244, Issue 1, August, pp. 3-50, 2019.

that "the plague was geographically vast and caused high mortality in some cases [...] it had a devastating short-term effect [...] however [...] the effects of the plague were neither uniform, nor so catastrophic as to cause substantial mortality at the societal level, not to mention the collapse of states or empires[33]". In a earlier article, Mordechai had asserted that the possible deaths for the city of Constantinople (with an estimated population of 500.000) during the first wave could have been "from 329,000 to 342,000 deaths with 67–71 days exceeding 100 mortalities per day[34]". Therefore, even though Procopius'assertion has been considered unreliable due to his criticism towards Justinian, as shown in the *Anecdota,* his estimates seem to be rcorrect. Mordechai and Eisenberg also confirm the version of John of Ephesus, who claims that the counting of the deaths was stopped when it reached 230.000, but he estimates that "300,000 people in Constantinople died during the first outbreak in the city"[35]. On the other hand, we can only speculate about the effects of the plague in the countryside[36]

However, Mordechai and Eisenberg's main assertion is that the Pandemic should not be considered a watershed event and they base their hypothesis on the analysis of the pollen, of the literary and non-literary sources (legislation, numismatics, papyri and archaeology)[37]. In their opinion, these analysis do not show sufficient proofs that the Justinian plague had major demographic effects; if this were the case, we could find clear evidence in legal texts or in the abandonment of exploited lands, measured by the palynological analysis. On the other hand, according to Mischa Meir, historian at University of Tubingen, Mordechai and Eisenberg's analysis is not sufficient to prove their conclusions. Meier, indeed, argues that the absence of an exhaustive account of the pestis must be explained by referring to the style and to the priorities of the historians.[38] According to the historian, we do not have to take in account the fact that "Procopius dedicates only 1.0% of his work to the plague[39]" or that "Gregory of Tours [...] dealt with the plague in 1.3% of his work[40]". Meier, on the other hand, stresses the fact that John Lidus announces a separate chapter on the plague in his *De magistribus* (which has gone lost), thus giving a great attention to the history (and possibly, to the effects) of the plague.

In addition, relatively to the purported absence of evidence of the plague in the legal texts, we can actually find some clues in the "7th edict of March 1, 542, [in which] reference is made quite bluntly to the encompassing presence of death in all places'[41]". Moreover, we should also consider the 122nd novel of 23 March 544 which had the aim of counteracting the price increases not only in the Capital, but also in the entire Roman East; furthermore, the so called "*epibolè*", the allocation

[33] *Ibidem.*

[34] White, L.A., Mordechai, L., *Modeling the Justinianic Plague: Comparing hypothesized transmission routes*, PLoS ONE 15,4, 2020.

[35] *Ibidem.*

[36] Meier, M., *The 'Justinianic Plague': An "Inconsequential Pandemic"? A Reply,* Medizinhistorisches Journal, Volume 55, 2, pp 172-199, 2020.

[37] Mordechai, L., Eisenberg, M., *Rejecting Catastrophe: The Case of the Justinianic Plague*, cit.

[38] Meier, M., *The 'Justinianic Plague': An "Inconsequential Pandemic"? A Reply,* cit.

[39] *Ibidem.*

[40] *Ibidem.*

[41] *Ibidem.*

of land to certain person who had to pay the tax burden lying on these lands, was probably due to the abandonment of land, as a result of the plague death toll[42].

As for the relative absence of references to the plague in the epitaphs, Meier argues that this is not a proof, since the plague is said to have mainly affected the poor, who could have not afforded an epitaph. Moreover, it must also be considered that people who died for the effects of the plague were predominantly buried in mass graves [43] and not in "normal" graves.

In response to the alleged absence of indications of losses made by Mordechai and his colleagues, Meir also include in his analysis the fact that "apparently, people in Egypt began to conclude long-term leases in the middle of the 6th century, which in turn reflects a change in the assessment of the value of agricultural land and may be related to the experiences during the plague"[44]. This would also reveal new insights in relation to the penetration of the pandemic in the countryside. Indeed, the effects of the plague in the countryside are also reflected in the abovementioned *epibolé* and in the description of Paul the Deacon of the Italian countryside.

Furthermore, in Meier's opinion, the palynological analysis should be embedded in "further source material", as it should regard lands that we are sure were actually used for agriculture. Lastly, in rejecting Mordechai's opinion, Meier argues that the plague was a major shock for the fact it was "the condensing moment of a series of events extending over several decades[45]". Contemporary sources, indeed, stress the fact that the period was also characterized by several natural events, not to mention the wars and the so called barbaric invasions[46]. This explains the words of Paul the Deacon, which describes abandoned villages and deserted fields as well as a huge silence among empty cities at the time of the pestilence of Italy[47]. Interestingly, this is the same description given by John of Ephesus.[48]

The sense of unpredictability present in the historians' description is also considered a proof of the huge demographic impact given by the pandemic: "John of Ephesus – Meir argues – wondered every morning whether he would live to see the evening"[49]; in his *Chronicles,* copied by Dionysus Telhmaharensins, he wrote:

> "One fell as he spoke, the other ran away; another died while eating; every man lost hope of living, and was afraid to go out, saying: "I shall perish in the middle of the house"[50].

In conclusion, we probably cannot precisely estimate the exact mortality rate of the first plague pandemic, but we should predominantly rely on the secondary effects to evaluate its demographic impact. Evidence seems to reveal it had a huge impact on

---

[42] Retief, F.P., *The epidemic of Justinian (AD 542): a prelude to the Middle Ages*, cit.

[43] Harbeck M, Seifert L, Hanasch S, Wagner DM, Birdsell D, et al., *Yersinia pestis DNA from Skeletal Remains from the 6th Century AD Reveals Insights into Justinianic Plague*, cit.

[44] Meier, M., *The 'Justinianic Plague': An "Inconsequential Pandemic"? A Reply,* cit.

[45] *Ibidem.*

[46] Retief, F.P., *The epidemic of Justinian (AD 542): a prelude to the Middle Ages*, cit.

[47] Paolo Diacono, *Historia Langobardorum,* Foro Barbarico, GBL Grande Biblioteca, ebook, 2020.

[48] Meier, M., *The 'Justinianic Plague': An "Inconsequential Pandemic"? A Reply,* cit.

[49] *Ibidem.*

[50] Pseudo-Dionysius of 'Tell-Mahre', *Chronicles*, Liverpool University Press, 1996, pp. 74-98.

the society of the time, both on the demographic and on the mentality point of view. New studies and statistics_argue that the proportion of the global population that died of the Justinian plague between 541 and 767 can be estimated at 33% and 40%, while the population decline in infected areas between 541 and 544 may have been around 20-25% and the population decline in infected regions between 541 and 700 may have been around 50-60%[51]. A section of this analysis will be therefore dedicated to the study of the socio-economic effects of the pandemic.

We will now focus on the geographical spread of the first plague pandemic.

1.1.2 The first plague pandemic across Europe

According to the chronicles of Michael the Syrian, who reports the testimony of John of Asia, the plague, after its outbreak in the Red Sea, spread to the rest of the world, that is to say, "to the regions of the West, which are called "upper", the peoples of the Romans, the Italians, the Gauls, and the Spanish"[52].

As far as Italy is concerned, we know that he the first epidemic occurred in 543[53]. However, the most im portant historical source that gives an extensive description of the phenomenon in Italy is Paul the Deacon, who records an outbreak in Liguria during 565[54]. After a brief description of the symptoms, the author of *Historia Langobardorum* reports the effects on the country:

> "Everywhere there were mourning and tears, and since word had spread that those who left their homes had a good chance of surviving, these were left empty, abandoned, only the dogs remained as guardians. The flocks remained alone, the sheep were left alone, without a shepherd to watch over them. Where villages and crowded fortified places could be seen before, there was now an absolute silence. Sons fled, leaving the corpses of their parents unburied; Parents fled, leaving their children in raging fever. Those who remained to bury his near relative, he remained himself unburied. Abandoned places returned to primordial silence, no voices in the fields, no whistling of shepherds; no dangers of wild beasts among the cattle; no harm to domestic fowls. The crops outliving the time of the harvest, awaited the reaper untouched; the vineyard with its fallen leaves and its shining grapes remained undisturbed while winter came on. Silence reigned supreme where before the trumpets of war and the clang of arms were heard; no travellers and not even brigands, yet there were corpses as far as the eye could reach. Pastoral places had been turned into sepulchres for men, and human habitations had become places of refuge for wild beasts.[55]"

This description is also coherent with the demographic analysis provided by historians Barbiera e Dalla Zanna[56]. Indeed, the authors of the research "Popula-

---

[51] O'Neill, A., *Estimated death toll of the Plague of Justinian 541-767*, Statista LUISS, April 2020.

[52] *Ibidem*.

[53] Little, L.K., *Life and afterlife of the First Plague Pandemic* in Plague and the End of Antiquity: The Pandemic of 541-750, cit.

[54] *Ibidem*.

[55] Paul the Deacon, *Historia Langobardorum,* Foro Barbarico, cit., translation of the Author.

[56] Barbiera, I., Dalla-Zuanna, G., *Population Dynamics in Italy in the Middle Ages: New Insights from Archaeological Findings, Population and development review* 35, 2, 367 – 389. 2009.

tion Dynamics in Italy in the Middle Ages: New Insights from Archaeological Findings", argue that the data obtained reveal a demographic pattern consistent with the outbreaks of the first plague pandemic. While experts agree on the fact that after the first outbreak "the number of inhabitants in Italy remained fairly constant for centuries[57]", there is evidence for an "intermittent" demographic regime, i.e. a series of mortality crisis fol lowed by a rapid increase in the young population[58]. Paul the Deacon claims that the last plague outbreak in Italy took place in the summer of 680[59], while we know that the final one was recorded in 749/50 in Sicily and Naples[60]; remarkably, according to the cited study, the population began to increase "not until the 9th century"[61]. On the other hand, we should also consider that the demographic crisis may have been a consequence of both the plague and the events of that period. At the time, Italy was crossed by the armies of both the Goths and the Byzantines, sent by the Emperor Justinian to reconquer IItaly. Furthermore, it was at the same period that the Lombards decided to move to Italy: according to Paul the Deacon, following their decision, "fire swords appeared in the night skies of Italy", presaging the arrival of this Germanic tribe. For two decades, the Italian territory passed alternately from the Goths to the Byzantines, who fought over territories already affected by the pandemic; finally, in 568, the Lombards moved in Italy. The end of their kingdom would then overlap with the last pandemic outbreak[62].

As for Rome, Paul the Deacon wrote that "so many people were dying that bodies were placed two by two on litters to be transported to tombs outside the city"[63]. As said, the last outbreak was recorded in Southern Italy in the mid-seventh century: it was one of the last outbreaks of the disease[64]. In concluding the analysis of the effects of the plague in Italy, it is interesting to dwell on a phrase of Paul the Deacon:

> "But this misfortune struck the barbarians within the Italian territory, up to the border with the Alemanni and the Bavarians.[65]

Until a few decades ago, it was doubted that the plague had actuallly crossed the Alps, but new discoveries have confirmed the assertion of Paul the Deacon[66] – thus,

---

[57] *Ibidem.*

[58] *Ibidem.*

[59] Little, L.K., *Life and afterlife of the First Plague Pandemic* in Plague and the End of Antiquity: The Pandemic of 541-750, cit.

[60] *Ibidem.*

[61] Barbiera, I., Dalla-Zuanna., G., *Population Dynamics in Italy in the Middle Ages: New Insights from Archaeological Findings*, cit.

[62] Little, L.K., *Life and afterlife of the First Plague Pandemic* in Plague and the End of Antiquity: The Pandemic of 541-750, cit.

[63] *Ibidem.*

[64] McCormick, *Toward a Molecular History of the Justinianic Pandemic*, in Plague and the end of antiquity, the pandemic of 541-750, cit.

[65] Paul the Deacon, *Historia Langobardorum*, Foro Barbarico, cit., translation of the Author.

[66] Harbeck M, Seifert L, Hänsch S, Wagner DM, Birdsell D, Parise KL, et al. Yersinia pestis DNA from Skeletal Remains from the 6th Century AD Reveals Insights into Justinianic Plague, cit.

probably his reliability in describing the effects of the disease in Italy. It is also important to point out – as we will see later – that the discovery made by the authors of the study *"Yersinia pestis DNA from Skeletal Remains from the 6th Century AD Reveals Insights into Justinianic plague"* is the only evidence of the presence of the plague north of the Italian mountain range[67]. This was possible thanks to the archaeological excavations carried out in the early-medieval cemetery of Ascheim, near Munich. As far as we know, Paul the Deacon is the only historical source who claim that the plague effectively crossed the Alps.

However, before focusing on the delicate issue of the presence of the plague pandemic in Northern Europe, we will analyse its effects in the territories of present-day France, Spain, Great Britain and Ireland. As for France, the main historical source is Gregory of Tours, who described the effects of the plague in his book *Historia Francorum.* The first mention of the plague is in the 31st chapter of the fifth book and is anticipated, as in the narration of its arrival in Rome, by several prodigies, like "three of four suns in the sky"[68] a "comet with a ray like a sword"[69] and fires in the sky[70]. Then, in the city of Clermont (Haute-Savoy),

> "the plague came, and such a carnage of the people took place through the whole district that the legions that fell could not be counted. For when sepulchres and gravestones failed, ten or more would be buried in a single trench. Three hundred dead bodies were counted one Sunday [...] A wound the shape of a serpent would appear on groin or armpit and the man would be so overcome by the poison as to die on the second or third day. Moreover, the power of the poison rendered the victim insensible. [...] At that time Lyons, Bourges, Cahors, and Dijon were seriously depopulated from this plague"[71]

We know that this happened in 563 because it is narrated after the so called "Tauredunum event", a tsunami triggered by a massive landslide in Lake Geneva. Again, after another prodigious event in Paris, during which "true blood fell from the clouds",

> "there was a great plague that year among the people. The sickness took various forms and was severe with pimples and tumours which brought death to many. Still many who were careful escaped. We heard that at Narbonne in that year the bubonic plague was very fatal, so that when a man was seized by it, he had no time to live."

Again, in the seventh book, chapter I, we are told that, approximately in the year 584,

> "the plague grew worse in Albi, and the greatest part of the people had now died and few of the citizens remained[72]"

[67] *Ibidem.*

[68] Gregory, Bishop of Tours, *History of the Franks*, translated by Ernest Brehaut, Fordham University online.

[69] *Ibidem.*

[70] *Ibidem.*

[71] *Ibidem.*

[72] *Ibidem.*

The last mention of the plague in France can be found in the ninth book, chapter XXI; interestingly, in the chronicles we find that king Childebert II refuses to send his armies against the Lombards "for a very severe plague is now wasting Italy". Later we know that the same king

> "was told at the time that Marseilles was suffering greatly from the bubonic plague and that the disease had spread swiftly as far as the village in the country of Lyons called Octavus. But the king [...] directed that nothing else than barley bread and clean water should be taken in the way of food and that all without intermission should keep watch."[73]

As for the city of Marseilles we are told that:

> "a ship from Spain [...] brought the seed of this disease. And many citizens bought various merchandise from her, and one household in which were eight souls was quickly left vacant, its inmates all dying of this plague. But the fire of the plague did not at once spread through all the houses, but after a definite time like a fire in standing grain it swept the whole city with the flame of disease. [...] The plague passed away in two months, and when the people, now reassured, had returned to the city the disease came on again and they who returned perished. Later on, the city was many times attacked by this death."[74]

This happened in 588[75], but the plague also returned three years after. As Gregory clearly states, the disease was brought through the trade routes up to the interior of the country; we know that Arles was struck by the pandemic in 551 and since the city is situated by the Mediterranean Sea and at the mouth of the Rhone river, it probably travelled along this river. Further evidence confirming that the spread of the plague followed the trade routes along French rivers is found in the necropolis of Sens, contemporary to the First plague Pandemic, which has been studied by historians Dominique Castex and Sacha Kacki[76].

On the other hand, Gregory of Tours also states that several parts of France were left out by the plague: that was the case of Clermont and the Auvergne[77]. According to Jean-Noel Biraben, cited by Dick Harrison, northern Gaul and the Midi were also unaffected by the plague[78]. This, according to Harrison, would confirm what previously stated, that the disease mainly spread along the trade routes.

Since we are told that the plague arrived in Marseille from Spain, we have a clue that is also hit the Iberian Peninsula. Unfortunately, this is only one of the few pieces of written evidence that we have[79]. This is a problem raised by both the historians

---

[73] *Ibidem.*

[74] *Ibidem.*

[75] McCormick, *Toward a Molecular History of the Justinianic Pandemic*, in Plague and the end of antiquity, the pandemic of 541-750, cit.

[76] Castex, D., Kacki, S., *Demographic Patterns Distinctive of Epidemic Cemeteries in Archaeological Samples*. Microbiology Spectrum, American Society for Microbiology, 2016.

[77] Sallares R., *Ecology, Evolution, and Epidemiology of Plague* in Plague and the end of antiquity, the pandemic of 541-750, cit.

[78] Harrison, D., *Plague, Settlement and Structural Change at the Dawn of the Middle Ages*, Scandia, Vol 59,1, 1993.

[79] Kulikowski, M., *Plague in Spanish Late Antiquity*, in Plague and the end of antiquity, the pandemic of 541-750, cit.

Micheal Kulikowski[80] and Jordina Carbonnell in two different studies[81]. The circumstances of the arrival of the disease in Spain are almost the same experienced by Italy; within two centuries, Spain went from being under the Romans, to the Visigoths, to the Alans and back to the Visigoths[82]. Moreover, during Justinian's reign, in 552, Spain was also invaded by the Byzantine Army, which established the province of *Spania*. At the time of the first outbreak of the Justinian plague in Spain – between 541 and 543[83] – the Iberian Peninsula was divided between the Visigoths' Kingdom – with Toledo as capital – the Kingdom of the Suebi – with Braga as capital – and the Byzantine province.

In this chaotic context, the pandemic was also spread by the Byzantine army, but it had already weakened the Kingdom of the Visigoths as well[84]. It had probably arrived in the main ports of Spain before the summer of 542 thanks to the close commercial ties with the Byzantine Greece[85], although we can only rely on the notes on the chronicles of Victor of Tunnuna, the so-called *Chronicles of Saragoza,* the only literary source for the initial outbreak of the disease[86]. In this text we find reference to an "inguinal plague" spreading "almost the whole of Hispania". To understand the lack of historical sources on the plague, we have to consider the social context of the Iberian Peninsula of the time. The post-Imperial Roman *Hispania* was, in fact, becoming a rural society, and was experiencing widespread violence, economic hardships and various diseases and epidemics, in addition to the Justinian plague[87]. These were aggravated by the common poor hygienic conditions and the economic hardship triggered by the events of the Migration Period. Therefore, historians can only depend on some hints rather than on clear evidence or historical sources. Even where we find massive burials, clear indication of an infectious disease, we can only hypothesise a link with the Justinian plague, since no paleogenetic studies have been carried out at these sites so far[88]. For instance, collective graves dating to the 6th century have been found in Valencia: no organization and no funerary rites has been detected there[89]. Moreover, as we find no evidence of violence and as the corps appear to have been burned or simply thrown into caves, we can conclude that this could be evidence of the presence of the Justinian plague. An extensive study to detect massive burials relating to the First plague Pandemic was done by Maria Bena-

---

[80] *Ibidem.*

[81] Sales-Carbonell, J., *Reaccions i estratègies davant l'arribada de la «pesta de Justinià» a la Península Ibèrica (segles VI-VII),* Revista d'igualada, 64-65, 2020.

[82] *Dall'Impero Romano a Carlo Magno,* La Storia, La Biblioteca di Repubblica, UTET.

[83] Kulikowski, M., Plague in Spanish Late Antiquity, in lague and the end of antiquity, the pandemic of 541-750, cit.

[84] Sales-Carbonell, J.,*Reaccions i estratègies davant l'arribada de la «pesta de Justinià» a la Península Ibèrica (segles VI-VII),* cit.

[85] Kulikowski, M., *Plague in Spanish Late Antiquity,* in Plague and the end of antiquity, the pandemic of 541-750, cit.

[86] Sales-Carbonell, J.,*Reaccions i estratègies davant l'arribada de la «pesta de Justinià» a la Península Ibèrica* (segles VI-VII), cit.

[87] *Ibidem.*

[88] *Ibidem.*

[89] Kulikowski, M., *Plague in Spanish Late Antiquity,* in Plague and the end of antiquity, the pandemic of 541-750, cit.

vides, History Graduate at University of Alcalà[90]. The fact that there are several improvised burials, such as caves and galleries, can be interpreted as a indicative of fear and oversaturation of burials. In this sense, the indications given by the Valencia council in 546 seem quite revealing: it is reported that in case of a sudden death, the tomb must be quickly sealed. Fifteen years later, another council in Braga ruled that bodies should not be buried in the walls of the basilicas.

As far as the historical sources are concerned, we can find a hint of the plague in the death of a man narrated in one chapter of the *Life of the Fathers of Mérida*, which recounts episodes from the 6th century; since the same chapter mentions "food shortages" "diseases" and "plague epidemics", we can argue that he probably died of bubonic plague, judging from his medical picture[91].

Another piece of evidence reported by Jordina Sales-Carbonell is an epigraph found in Cordoba dating back to the year 609, which describes a person who died of a plague in the groin[92].

Given the widespread evidence, Carbonell argues that by the seventh century, the plague was already endemic[93]. However, according to Kulikowski, the continuous outbreaks could have not been possible without constant reintroduction of the disease from outside the Iberian Peninsula[94]. This is because, according to the historian, even at its peak, the population was "not large enough to sustain the disease endemically"[95]. In fact, constant reemissions may have been possible due to the continuous invasions that struck the former Roman province of *Hispania*. In 711, the armies of Tariq ibn Ziyad invaded Spain, establishing the Umayyad Caliphate. Two year before, according to the *Akhbar Majmu'a,* an Arabic source, between 707 and 709 half of the Spanish population died in consequence of famine and the plague[96]. However, we have little information about subsequent outbreaks of the disease after the invasion of 711[97].

In conclusion, we can state that, even though we cannot rely on a precise description of the first plague pandemic in Spain, evidence show that it had profound effects on the society of the time: however, we will dwell on this specific point later. Nonetheless, it is important to stress that the historical sources analysed so far are coherent with the description given by the contemporary ones in Italy and France. At the same time, it is also crucial to emphasize that the demographic decline witnessed by the abandonment of rural settlements from the beginning of the sixth century can also be attributed to the chaotic conditions of the Iberian Peninsula. The Justinian plague, therefore, aggravated the condition of a society already hit by the dramatic changes following the collapse of the Roman Empire.

---

[90] Benavides, M., *Evidencias arqueológicas de la peste justinianea en Hispania*, Arqueología y Prehistoria del Interior Peninsular, 8, 2019.

[91] Sales-Carbonell, J.,*Reaccions i estratègies davant l'arribada de la «pesta de Justinià» a la Península Ibèrica (segles VI-VII)*, cit.

[92] *Ibidem*.

[93] *Ibidem*.

[94] Kulikowski, M., *Plague in Spanish Late Antiquity*, in Plague and the end of antiquity, the pandemic of 541-750, cit.

[95] *Ibidem*.

[96] *Ibidem*.

[97] Sales-Carbonell, J.,*Reaccions i estratègies davant l'arribada de la «pesta de Justinià» a la Península Ibèrica (segles VI-VII)*, cit.

Epochal changes make it difficult to detect the effects or even the arrival of the Justinian plague in the British Isles as well. This period, indeed, was dense of transformations and tensions: the end of the Roman rule in Britain, the simultaneous raids of Saxons, Picts and Angles and the establishment of a multitude of little kingdoms scattered across the island (as narrated by Gildas in his *De Excidio et Conquestu Britannia)*.

However, among the few literary sources we can find the work of the said Gildas, probably a member of a Breton noble family: once he become a monk, he travelled to Ireland and to Rome, and he founded a monastery in Rhuys, Britanny[98]. There, he wrote his most famous work, *De Excidio et Conquestu Britannia* and he probably died around 570, in his seventies. In his work we can probably find a reference to the bubonic disease that was spreading in Europe: indeed, in Part I, chapter 22, he writes about a *"famosa peste"*, that is to say a "well known plague"[99]:

> "a deadly pestilence came upon the unwise people which, in a short time, without any sword, brought down such a number of them that the livings were unable to bury the dead"[100]

The use of the adjective "well known" seems to suggest that the disease was widespread in the country: interestingly, Gildas travelled to Ireland, where we can find a possible reference to the arrival of the plague in the Annals of Ulster of 545[101]: *"Mortalitas prima que dicitur blefed in qua Mo-Bi Clâranech obit* [that is to say] The first mortality which is called bléfed, in which Mo-Bi Clârai died"[102]. In addition, in the *Annals of Clonmacnoise* we are told that *"The[re] was a great mortality w[hi]ch was called Irish Blefeth, of which Disease Clarineagh al[ia]s Berchann, who is supposed to be called in English Merlyn"*[103].

We cannot determine whether the aforementioned "disease" was actually the Justinian plague, although the timing seems to suggest it. It may indeed be possible, since the plague appeared in 541 in Egypt and there were close connections between the British Isles and the Mediterranean world at the time; furthermore, there is evidence that this trade also involved corn and cloth, commodities associated to rats and fleas [104]. Moreover, in the Annals of the Four Masters we read that:

> "[in] The Age five hundred and forty-three. The fifth year of Diarmait. There was an extraordinary universal plague through the world, which swept away the noblest third part of the human race".

---

[98] *Gildas il Saggio*, Enciclopedia Treccani online.

[99] Gildas, *The ruin of Britain, fragments from lost letters, the Penitential, Together with The lorica of Gidas*. Edited for the Hon. Society of Cymmrodorion by Hugh Williams, Published by the the Hon. Society of Cymmrodorion by David Nutt, 1899.

[100] *Ibidem*.

[101] Haley, G.C., *Tamlachta: The Map of Plague Burials and Some Implications for Early Irish History*, Proceedings of the Harvard Celtic Colloquium ,22, pp. 96-140, 2002.

[102] *Ibidem*.

[103] *Ibidem*.

[104] Maddicot, J.R., *Plague in seventh-century England*, Past & Present, Aug., 1997, No. 156 (Aug., 1997), pp. 7-54.

The reference to the "world" and the "third part of the human race" is effectively consistent with the spread of the plague and the mortality rate we found in other parts of Europe. Interestingly, on the other part of the Sea, in Wales, the King Maelgwn died during the "great mortality of 547", as reported by the *Annales Cambriae*[105]. Another reference to the "great mortality" is made in the Annals of Ulster[106]

Moreover, the historian William MacArtur reports that "The pestilence that is called *samtrusg*"[107], recorded in 554, should be associated to something affecting the skin, as the name seems to suggest[108]: however, given its degree of infectivity implicit in the word "pestilence", it cannot be identified with leprosy, as suggested by the Annals of Ulster.

We have to wait until the chronicles of Beda – who lived between 673 and 735 – for a literary evidence of what could have possibly been the arrival of the plague in England. It is important to stress that, unlike Procopius and his contemporary historians, Beda had not probably read Thucydides' description of the pestilence in Athens and when the plague arrived in the monastery where he lived, he was only a teenager[109]. Nonetheless, his description of a disease is precise:

> In the same year 664, there happened an eclipse of the sun, on the third day of May, about the tenth hour of the day. In the same year, a sudden pestilence depopulated first the southern parts of Britain, and afterwards attacking the province of the Northumbrians, ravaged the country far and near, and destroyed a great multitude of men. [...] Moreover, this plague prevailed no less disastrously in the island of Ireland."[110]

The time of the year is consistent with the Justinian plague: indeed, it usually appears in late spring and becomes more virulent in summer, when fleas reproduce[111]. Interestingly, in the same year 665, the Annals of Tigernach (Ireland) record a "great mortality"[112].

After the epidemic of 664, Bede records other outbreaks: according to John Maddicot there was one in the monastery of Deiran between 666 and 669, after which the pestilence headed south[113]. In chapter XXXIII of *Vita St. Cuthberti,* Bede states that:

> "at the same time the plague made great ravages in those parts, so that there were scarcely any inhabitants left in villages and places which had been thickly populated, and some towns were wholly deserted."[114]

---

[105] *Ibidem.*

[106] *Ibidem.*

[107] MacArthur, W., *The Identification of Some Pestilences Recorded in the Irish Annals*, Irish Historical Studies, 6, 23, 169-188, 1949.

[108] *Ibidem.*

[109] Maddicot, J.R., *Plague in seventh-century England,* cit.

[110] Sellar, A.M., *Bede's Ecclesiastical History of England, A Revised Translation, With Introduction, Life, and Notes*, Oxford, London George Bell and Sons, 1907.

[111] Maddicot, J.R., *Plague in seventh-century England,* cit.

[112] Crawford, M., *Famine and Disease in Ireland,* edited by A. Clarckson and Margaret Crawford, Tayor and Francis Ltd, 2005.

[113] Maddicot, J.R., Plague in seventh-century England, Past & Present, cit.

[114] Seller, A.M., Bede's Ecclesiastical History of England, A Revised Translation, With Introduction, Life, and Notes, cit.

According to William MacArthur, this outburst coincided with the so called *Mortalitas Puerorum,* the mortality of children in Ireland, which is called "the third mortality" in the Annals of Inisfallen[115]: this could be a reference to the fact that it was the third outburst of the same pestilence.

It has been said that the brevity of such information provided by the chronicles of the time might suggest that the Justinian plague never reached Ireland or that the population density was not sufficient to spread the disease[116]; this is why authors like Gene Haley are trying to find evidence of the plague in the *tamlachta,* that is to say in the mass burials found in Ireland. Moreover, on the issue of population density, Haley argues that at least in some locations, it may "have been higher than is usually envisioned"[117]. Findings of exotic pottery and jewellery and metallurgy in areas where there are several mass burials, seem to confirm the previously hypothesised link between trade and diffusion of the plague. In fact, we have to say that the only clear evidence that the plague ever reached the British Isles was found in a mass burial in Edix Hill, near Cambridge. New research and analysis of the people buried in the *tamlachta* could possibly reveal new insights on the spread of the Justinian plague[118].

Moreover, many authors argue that there could be a link between the possible arrival of the plague in Ireland and the spread of the so called "ringforts"[119]; remarkably, this kind of fortification was extremely rare in the Irish Iron Age and began to reappear during the fourth and the fifth century[120]. Unlike the rest of Europe, there were no social tensions and invasions in Ireland; the sudden and large construction of fortifications has been therefore interpreted as a response or as an attempt to protect from a mysterious and widespread mortiferous disease[121]. Others have also suggested that these buildings may have also beenr a way to demonstrate a certain status or the beginning of a social reorganization following the death of a great part of the population[122]. Indeed, no evidence of change of warfare have been detected so far. This interpretation is also consistent with the structural abandonment of settlements witnessed in the seventh-century England by Christopher Taylor, as reported by Maddicot[123] and testified by Bede. However, this has also been interpreted as evidence of demographic decline following the pandemic[124].

---

[115] MacArthur, W., The Identification of Some Pestilences Recorded in the Irish Annals.

[116] Haley, G.C., Tamlachta: The Map of Plague Burials and Some Implications for Early Irish History, cit.

[117] *Ibidem.*

[118] Keller, M., Spyrou, M.A., et al., Ancient Yersinia pestis genomes from across Western Europe reveal early diversification during the First Pandemic (541–750), cit.

[119] See Haley, G.C., Tamlachta: The Map of Plague Burials and Some Implications for Early Irish History, cit and Lynn, C.J., *Settlement and Disease: a plague on your raths* in *Archaeology Ireland*, 19,4, 14-7, 2005.

[120] Haley, G.C., Tamlachta: The Map of Plague Burials and Some Implications for Early Irish History, cit.

[121] *Ibidem.*

[122] Lynn, C.J., *Settlement and Disease: a plague on your raths*, cit.

[123] Maddicot, J.R., *Plague in seventh-century England*, cit.

[124] *Ibidem.*

1.1.3 What about Northern Europe?

A separate chapter is needed to address the issue of the spread of the Justinian plague in Scandinavia. On this subject, indeed, there is no broad agreement[125], but several hypotheses have been made so far. First of all, while for the rest of Europe we can speculate on the presence of the plague by analysing several hints given by the chronicles, this is not the case in Northern Europe: here there is a complete lack of written sources[126]. Therefore, the approach of historians has been to look for evidence of possible outbreaks of the Justinian plague, such as mass burials and depopulated settlements, as in the rest of Europe.

Interestingly, there is evidence of population decline throughout northern Europe in the years following the outbreak of the Justinian plague: according to the historian Daniel Löwenborg, all over Scandinavia "there are indications of a dramatic and perhaps sudden societal, economical, ecological as well as demographic downturn, or at least of broad change. This change has led Swedish archaeologists to consider this period as a transition period between the Early and the Late Iron Age. The turning point is usually set at year CE 550[127]". For instance, the two historians Graslund and Price report that in Southern Norway there was a decrease in the number of burials around the middle of the 6th century[128], while Daniel Löwenborg states that the clearest evidence of a general abandonment can be found in central Sweden, in the village of Hälsingland, where palynological analysis has shown a drastic decline in pollen and cereals and other vegetations related to human activities[129]. Similar observations have been made relatively to the islands of Gotland – where signs of abandonment of at least 1900 structures are reported – and Oland, where in addition to a structural abandonment, some bodies seem also appear to have been burnt[130]. Moreover, fortified settlements were also peculiar of that period, thus confirming that social tensions were common in Europe during the Migration Period[131]. Moreover, a shift in burial customs in Scandinavia seems to be part of a general European modification in funeral habits[132]. New research should therefore investigate the possibility that this may be related to the spread of the plague. In Finland, for instance, a general low burial continuity between the 6th and the 7th century – particularly evident in Ostrobothnia – has been interpreted as evi-

---

[125] *Plague and the End of the Antiquity, The pandemic of 541-750,* Edited by Lester K. Little, Cambridge University Press in association with The American Academy in Rome, 2007.

[126] Seger, T., *The plague of Justinian and other scourges : an analysis of the anomalies in the development of the iron age population in Finland,* Journal of Swedish Antiquities, 1982.

[127] Löwenborg, D. *An Iron Age Shock Doctrine – Did the AD 536-7 event trigger large-scale social changes in the Mälaren valley area?,* Journal of Archaeology and Ancient History, 4, 1-29, 2012.

[128] Graslun, B., Price, N., *Twilight of the gods? The 'dust veil event' of AD 536 in critical perspective,* Antiquity, 86(332), 428-443, 2012.

[129] Löwenborg, D. *An Iron Age Shock Doctrine – Did the AD 536-7 event trigger large-scale social changes in the Mälaren valley area?,* cit.

[130] Flink, G., *Ölands stensträngsområden och den justinianska pesten,* Nordic Journal of Settlement History and Built Heritage, 11, 21–135, 1986.

[131] Löwenborg, D. *An Iron Age Shock Doctrine – Did the AD 536-7 event trigger large-scale social changes in the Mälaren valley area?,* cit.

[132] *Ibidem.*

dence of the presence of the plague, also considering the two mass burials found in the Ostrobothnian waters, containing at least 60 individuals.[133]

On the other hand, critics of this hypothesis argue that the species so far considered the main vector of the plague, *Rattus rattus* (grey mice), was not present in Scandinavia until around 1050 CE[134]; indeed, so far sporadic findings have only been made in Poland, (dated to the eight century) and near Berlin, at Waltersdorf[135] (Roman period) and except for a few discoveries made by the border of the Roman Empire in modern day Hungary[136], grey mice do not appear to have crossed the former *Limes* of the Roman empire; at least not until the 10th century. According to paleobiologists Audoin-Rouzeau and Vigne the grey mouse is a "sedentary rodent"[137] and only moves when transported by the humans. This would confirm the spread of the Justinian plague as limited to the nearabout of commercial roads, explaining why some cities were not affected. However, new research and discoveries are undermining this interpretation.

While we are sure that the Black Death spread all over Europe (apart from some defined areas), according to Zoology Professor David Davis, we are not sure that this was possible thanks to the action of *Rattus rattus*: rapidity of spread could have also been possible through contact or fleas[138]. This is also the thesis of researchers Anne Karin Hufthammer and Lars Walløe, who argue that there are no findings of black rats in Norwegian rural areas in Medieval times and they are limited even in harbour towns[139]. According to the research, this is also true for the other Nordic countries. Therefore, considering *Rattus Rattus* as the only intermediate host of *Yersinia Pestis* would be problematic, given the certain effects of the Second plague Pandemic in Scandinavia[140] . This is not to say that there is not a link between plague and *Rattus Rattus*[141]; nonetheless, their presence seems not strictly linked to the one of bubonic plague. Another clear example is provided by the bubonic plague of London, where no rats are mentioned[142]. Moreover,

---

[133] Seger, T., *The plague of Justinian and other scourges : an analysis of the anomalies in the development of the iron age population in Finland*, cit.

[134] Davis, E.D., *The Scarcity of Rats and the Black Death*: An Ecological History, The Journal of Interdisciplinary History, 16, 3, 455-470, 1986.

[135] Audoin-Rouzeau, F., Vigne, J., *The black rat (Rattus rattus) in Roman and Medieval Europe: The commercial roads and the spread of plague*, Anthropozoologica, 1997.

[136] Kovács, Z.E., *Dispersal history of an invasive rodent in Hungary – subfossil finds of Rattus Rattus*, Acta Zoologica Academiae Scientiarum Hungaricae 58, 4, 379–394, 2012.

[137] Audoin-Rouzeau, F., Vigne, J., *The black rat (Rattus rattus) in Roman and Medieval Europe: The commercial roads and the spread of plague*, cit.

[138] Davis, D.E., *The Scarcity of Rats and the Black Death: An Ecological History*, cit.

[139] Hufthammer, A.K.,Walløe, L., *Rats cannot have been intermediate hosts for Yersinia pestis during medieval plague epidemics in Northern Europe*, Journal of Archaeological Science, Elsevier, Volume 40, 4,1752-1759, 2013.

[140] According to Harrison, author of *Stora döden: den värsta katastrof som drabbat Europa*, the disease killed as much as 60-65% of the population in the Nordic country and was brought by ship from England.

[141] *Hufthammer,* A.K.,Walløe, *L., Rats cannot have been intermediate hosts for Yersinia pestis during medieval plague epidemics in Northern Europe,* cit.

[142] *Ibidem.*

even when mice's presence is certain, studies are not sure about the effective link between rodents and plague: in fact, the territories of groups of rats are larger than a house and this does not fit with circumstances in which plague was confined to a single house[143]. Hufthammer and Walløe's conclusion is worth reporting as it probably solves the long-standing dispute regarding the spread of the Justinian plague:

> "Both the epidemics during the Justinian pandemic (542-767) and the European epidemics from 1347 onwards were plague epidemics, in other words caused by *Y. pestis*. During these epidemics in Europe, plague was transmitted directly from human-to-human by an insect ectoparasite vector, without a mammalian vector such as the black rat. The only potential candidates that were widely distributed in large numbers in all European countries appear to have been the human flea *P. irritans* and the human louse *P. humanus*. There were probably very large numbers of fleas and lice in people's clothes and bedding in the Middle Ages and early modern times. Over longer distances, plague was carried by people making journeys or in the goods they transported. Infected human fleas can survive for long periods without feeding, and could therefore have been trans-ported in clothing, wool and many other types of goods. This trans-mission model, unlike the rat model, can also explain the rapid spread of plague epidemics. It also explains why all members of one household in a town might become plague victims while neighbouring households escaped. Many such cases were observed, and they cannot be explained by the rat model"[144]

Interestingly, a 2014 study confirmed that *Pulex irritans* can transmit *Yersinia Pestis* between humans[145]. Moreover, we also have to consider that even though Northern Europe was outside the former borders of the Roman Empire, it was fully included in the trade routes of the time[146]. There were three main commercial itineraries connecting Southern and Central Europe to Scandinavia: the eastern one, which connected the Eastern Mediterranean to the Baltic via the Russian steppes; the central one, linking Germany to Northern Europe, where traders mainly transported fur and gold; the western one, where goods travelled from Gaul across the North Sea[147].

In conclusion, another recent – and unexpected[148] – discovery seem to confirm Hufthammer and Walløe's thesis: a 2018 study revealed the presence of *Yersinia Pestis* in two individuals in a Neolithic settlement in Sweden. The genomic analysis carried out by the scientists demonstrated that the bacterium's strain predated all the "modern and ancient strains of this pathogen" and that the results of the re-

---

[143] Davis, D., The characteristics of rat populations, The Quarterly Review of Biology, pp. 373-401, 1958 in Hufthammer, A.K.,Walløe, L., *Rats cannot have been intermediate hosts for Yersinia pestis during medieval plague epidemics in Northern Europe*, cit.

[144] Hufthammer, A.K.,Walløe, L., *Rats cannot have been intermediate hosts for Yersinia pestis during medieval plague epidemics in Northern Europe*, cit.

[145] Ratovonjato, J., Rajerison, M., Rahelinirina, S., Boyer, S., *Yersinia pestis in Pulex irritans Fleas during Plague Outbreak*, Madagascar, Emerg Infect Dis., 20,8, 1414-1415, 2014.

[146] Jankuhn, H., *Trade and Settlement in Central and Northern Europe up to and during the Viking Period*, The Journal of the Royal Society of Antiquaries of Ireland , 112, 18-50, 1982.

[147] *Ibidem*.

[148] Zhang, S., *An Ancient Case of the Plague Could Rewrite History*, The Atlantic 6 December 2018.

search support the possibility of an "epidemic of plague" [149]. This means that the bacterium spread throughout Eurasia during the Neolithic decline[150]. Moreover, since no *Rattus Rattus* skeletons have been found earlier than Viking Age in Scandinavia, this proves that *Yersinia Pestis* may have been present anyway. Further studies on the skeleton remains dating to the Migration Period are therefore needed.

1.1.4 The socio-economic effects of the Justinian plague

We already dealt with the descriptions given by the historians about the abandonment of the territories: these images give the idea of a widely depopulated Europe but also raise the question of the long-term effects of the plague on a society. The European society of the time, indeed, was already facing historical changes such as the end of the Roman rule, the collapse of written knowledge and the incoming migrations of Northern European tribes: we have, therefore, to investigate and classify the effects of the plague.

In his essay *The plague pandemic and the Slavic expansion in the 6th-8th centuries,* Arkadiusz Sołtysiak, University of Warsaw, divided the effects of the plague into "physical" and "pychological" ones, arguing that we could identify primary, secondary and ternary effects. We may therefore identify different impacts from the point of view of the individuals, as well as on the demographic, economic, political, and social[151].

We will therefore follow this approach, looking for evidence in historical sources as well as in the archaeological findings.

First of all, we can investigate the effect on individuals: we have already mention that John of Ephesus used to say that he wondered whether he would live to see the evening[152] . Pseudo-Dionisius, for instance, tells of a man who wore a tablet with his names on it so that, if he died, his family would know[153]. These two narrations give the idea of a common sense of uncertainty. Moreover, Pseudo-Dionisius gives a precise description of the overall individual effects too, arguing that the plague's impact came:

> "over corpses which split open and rotted on the streets with nobody to bury (them); over houses large and small, beautiful and desirable which suddenly became tombs for their; inhabitants and in which servants and masters at the same time suddenly fell (dead), mingling their rottenness together in their bedrooms, and not one of them escaped who might remove their corpses out from within the house; over others who perished falling in the streets to become a terrible and shocking spectacle for those who saw them, as their bellies were swollen and their mouths wide open, throwing up pus like torrents, their eyes inflamed and their hands stretched out upward, and (over) the corpses rotting and lying on comers and streets, with nobody to bury (them); over ships in the midst of the sea

---

[149] Rascovan, N., Sjögren, K., et al., *Emergence and Spread of Basal Lineages of Yersinia pestis during the Neolithic decline*, Cell, Volume 176, 1-2, 295-305, 2019.

[150] *Ibidem.*

[151] Sołtysiak, A., *The plague pandemic and the Slavic expansion in the 6th-8th centuries,* Archaeologia Polona, Institute of Archaeology and Ethnology Polish Academy of Sciences, 44, 2006.

[152] Meier, M., *The 'Justinianic Plague': An "Inconsequential Pandemic"? A Reply,* cit.

[153] Pseudo-Dionysius of 'Tell-Mahre', Chronicles, cit.

[which] became tombs for their captains and they continued adrift on the waves carrying the corpses of their owners; over other (ships) which arrived in harbours, were moored by their owners, and remained (so), never to be untied by them again; over palaces which groaned one to the other; over bridal chambers where the brides were adorned (in finery), but all of a sudden there were just lifeless and fearsome corpses; over virgins which (had been) guarded in bedchambers and (now) there was nobody to carry them from (these) bedchambers to the tombs; over highways which became deserted; over roads (on) which (the traffic) was interrupted; over villages whose inhabitants perished all together; over many things of this kind, which defeat all who have the power of speech in (their skill with) words and stories."[154]

This description effectively describes the impact of the plague on every member of the society. Moreover, as there was no more control,

"If somebody said that in that shop there were only 1000 pounds of gold and silver, it would still be (as if) nothing. So [...] people entered and took freely. They looked and were astonished. They cast their eyes on many things, especially on gold, and took (it) and wished to leave"[155]

This is something that is also reported by Procopius, although in his description we see how people changed their conduct, many forgetting their membership of a certain faction, others ust starting to behave as they already had once the plague had passed:

"At that time all the customary rites of burial were overlooked. For the dead were not carried out escorted by a procession in the customary manner, nor were the usual chants sung over them, but it was sufficient if one carried on his shoulders the body of one of the dead to the parts of the city which bordered on the sea and flung him down; and there the corpses would be thrown upon skiffs in a heap, to be conveyed wherever it might chance. At that time, too, those of the population who had formerly been members of the factions laid aside their mutual enmity and in common they attended to the burial rites of the dead, and they carried with their own hands the bodies of those who were no connections of theirs and buried them."[156]

These descriptions not only make us imagine how the individual microcosm changed, but also us an idea of how the macrocosm was affected; Ps. Dyonisius, in fact, makes reference to deserted "ships", "highways", "roads" and "villages", as well as the problem of the burials. The response of the state is narrated by John of Ephesus, copied from Ps. Dyonisius:

Also "the empire was sitting in sorrow", as it is written, for (the authorities) learned that the hands of the people who were bringing out the corpses grew weak because they also became fewer and (began to) disappear. The city stank with corpses as there were neither litters nor diggers and the corpses were heaped up in the streets. Thus when the merciful emperor, in whose days these things took place, learned of it, he stirred himself up with zeal and showed diligence, giving orders for 600 litters to be produced. He appoint-

---

154 Ps.Dyonisius in Rosen, W., Justinian's Flea, Plague, Empire, and the Birth of Europe, cit.

155 John of Ephesus in Ps. Dyonisius of Tel-Mahre, *Chronicle, Part III,* translated with notes and introduction by Witold Witakowski, Liverpool University Press, 1996, p.74-98.

156 Procopius, Book II. 22., *translation of. H. B. Dewing,* cit.

> ed a man, his referendarius, whose name was Theodore, who was also zealous in good deeds, and gave him instructions to take and spend as much gold as should be necessary for supervising these matters and for encouraging people with great gifts not to be negligent but to dig large ditches and to fill them by piling up the corpses. This man proceeded with application. [...] He took along many people, gave them much gold and had very large pits dug, in every one of which 70, 000 (corpses) were put. [...] He himself was ordered to fill every grave he could find, to whomsoever it might belong. Thus, by his application the city was gradually rid of the corpses. Everyone who had many corpses (to be buried) went to inform him and he would have them removed."[157]

The secondary obvious effect of these deaths was the demographical decline, which we had analysed yet. This demographic decline is reflected for instance in the lack of people to bury the dead. However, as we have already seen that in their chronicles, both Paul the Deacon and John of Ephesus claim that fields were left deserted[158], the demographic decline therefore also had important repercussions on the economy. Increasing evidence, indeed, is confirming that fields were actually abandoned: this is proved by the palynological analysis and by various research which reveal a growth of the forests in that period[159]. Moreover, this is also confirmed by the abovementioned tax, the *epibolè,* which reallocated fiscal duties which were present on deserted fields[160]. Interestingly, the historian Peter Sarris argues that the demands of agricultural workers were met with an increase in wages[161]: on the other hand, an edict issued in the April of 545 – which was to take effect throughout the Eastern Roman Empire – stipulated that "construction, agricultural and other group of "workers"[162] should not receive a wage increase; moreover, Sarris reports that Justinian complained that, following the arrival of the plague, "tradesmen, artisans and agricultural workers" demanded "double or triple the prices and wages"[163]. This definitely confirms the general lack of labour at the time. As we can see from the edict, trade was affected too; in fact, there is evidence for a general decrease in trade all over Europe: in Spain, trade between the peninsula and Africa diminished after 560[164], as well as the one with the Eastern Empire[165]; in France, the towns along the Bay of Biscay experienced a decline in trade in the same period[166]; in Britain, from 600 onwards, there is evidence of imitation Byzantine gold *solidi*[167], the currency of the Empire, probably due to a local shortage of it; in the islands of

[157] John of Ephesus in Ps. Dyonisius of Tel-Mahre, *Chronicle, Part III,* cit.

[158] See page 17.

[159] Graslund, B., Price, N., *Twilight of the Gods? The 'dust veil event' of AD 536 in critical perspective,* cit; Sallares,R., Ecology, Evolution, and Epidemiology of Plague, cit; Harrison, D., Plague, Settlement and Structural Change at the Dawn of the Middle Ages, cit.

[160] Sarris, P., *Bubonic plague in Byzantium,* in Little, K.L. (ed), Plague and the end of antiquity, the pandemic of 541-750, cit.

[161] *Ibidem.*

[162] *Ibidem.*

[163] *Ibidem.*

[164] Kulikowski, M., *Plague in Spanish Late Antiquity,* cit.

[165] *Idibem.*

[166] *Ibidem.*

[167] Jellema, D., *Frisian trade in the Dark Ages, Speculum,* The University of Chicago Press, 30, 1, 15-36, 1955.

Gotland, Oland and Bornholm in the Baltic sea, the large amount of gold Byzantine *solidi* dating to the 6th century, "breaks off [...], after mid-century"[168]; actually, the historian Jankuhn confirms that the route which connected southern to northern Europe is interrupted around 560[169]. The reduction of trade undoubtedly led to a general decrease in wealth: this is showed again by Peter Sarris, who observes that the "fraility of the state finances is evident from the copper coinage [whose] weight [...] declined by some 23%"[170]. Moreover, the author also reports that Procopius describes a reduction in the State's expenditure on pensions and a reduction of the military budget. We will deal on the probable effects of this cut later. However, we might hypothesise that famine and shortages followed the decrease in trade and in the agricultural production.

These economic changes would in turn have repercussion on social organization: for instance, Dick Harrison, referring to the work of the historian Montanari, argues that the "agricultural economy of the Roman times [reverted to] a more sylvan-pastoral structure"[171]. Paul the Deacon argued that in Italy many had fled both the cities and the countryside[172]; we have already seen evidence of depopulation in many European settlements, from Ostrobothnia[173] to Britain[174] and Spain[175], but it is worth recalling it in order to suggest a possible reshape of the way of living. Indeed, we found that, along with evidence of depopulation, we witnessed a general "fortification" of settlements: this was true for the ringforts in Ireland, the fortifications in the Baltic as well as for the "hilltop settlements" in Southern Italy[176] and the one of the Lombards in Northern Italy[177]. People who were not affected by the plague nevertheless decided to migrate in response to the disease, looking for new accommodations: this is evident, for instance, in Britain, as Maddicot suggests[178]. The countryside's landscape changed: the roman *villa,* a rural settlement, gave way to the better defendable villages in the mountains or in the Alpine valleys[179]: as Harrison claims, this was true from Gaul to Bulgaria[180]. We can however argue that this was also due to the turbulence of the Migration Period and the need to protect from a mysterious disease linked to trade and contact.

The constant depopulation of the countryside had also important geopolitical effects: we saw earlier seen how military budget of the state was reduced: together with the depopulation of the countryside, this would have provided a strong incentive for

[168] *ibidem.*

[169] Jankuhn, H., *Trade and Settlement in Central and Northern Europe up to and during the Viking Period,* cit.

[170] Sarris, P., *Bubonic plague in Byzantium.*

[171] Harrison, D., *Plague, Settlement and Structural Change at the Dawn of the Middle Ages,* cit.

[172] Paul the Deacon, *Historia Langobardorum,* Foro Barbarico, cit.

[173] Seger, T., *The plague of Justinian and other scourges : an analysis of the anomalies in the development of the iron age population in Finland,* cit.

[174] Maddicot, J.R., *Plague in seventh-century England,* cit.

[175] *Ibidem.*

[176] Harrison, D., *Plague, Settlement and Structural Change at the Dawn of the Middle Ages,* cit.

[177] Interview with Professor Lucarelli, University of Oslo.

[178] Maddicot, J.R., Plague in seventh-century England, cit.

[179] Harrison, D., *Plague, Settlement and Structural Change at the Dawn of the Middle Ages,* cit.

[180] *Ibidem.*

bordering population to settle in these inhabited territories. This is, at least, the thesis of several historians, who link the expansion of the Slavs in the Balkans[181] and that of the Arabs[182] to the plague pandemic. Nonetheless, as we are focusing on Europe, we will only analyse the migration of the Slavs.

According to Sołtysiak, the Slavic expansion could be linked to the disinterest in defending the northern frontier or to the agricultural over-taxation but not to the spread of the plague, despite the coincidence between the first Slavic raids and the outbreak of the plague. However, we might add that both the over-taxation and the disinterest for the northern frontier could be interpreted as caused by the plague: indeed, as already seen, the loss of revenues due to the depopulation of the countryside made the Emperor raise the *epibolè* tax; on the other hand, the same loss of revenues led to a reduction of the military expenditure, with important consequences at the border in an already turbulent period. Slavs had already entered the *Limes,* though it is only in the 40s that they besieged several strongholds in *Illyricum* and *Thracia*[183]*,* as far as Epidamnus, in modern day Albania[184]. Procopius adds that these once "reputed strong places" were now "quite undefended"[185]. In the following years, Byzantium lost control of the Balkans, while the Slavs took control of the villages[186]. On the other hand, according to Procopius, many farmers, desperate because of the *epibolé,* even fled into the lands of Barbarians[187]. Justinian too, according to the historian, not being able to adequately pay the Army in the Balkans, preferred to pay the Barbarians also in order to make them his allies[188].

However, in order to restore the control of these lands, Justinian's successors began to resettle and colonise the abandoned territories of the Empire: the Slavs were thus deported in Peloponnesus, Bithinia and Cyprus, sometimes even voluntarily[189]. They were also recruited in the army and the situation seemed to improve, as evidenced by Harrison[190]. Anyway, we will further deal with the migration of the Slavs in the following chapter.

Lastly, the plague left also psychological effects and changed the habits of the people of the time; Michael Morony reports the testimony of John of Ephesus in which he argues that "everyone was in perturbation and doubt and confusion."[191] People fled and sought refuge in other cities, probably spreading the plague[192], but, according

[181] Sołtysiak, A., The plague pandemic and the Slavic expansion in the 6th-8th centuries, cit.

[182] Little, L.K., Life and Afterlife of the First Plague Pandemic, cit

[183] Sołtysiak, A., The plague pandemic and the Slavic expansion in the 6th-8th centuries, cit.

[184] Procopius, *History of the Wars 7 Vols., trans. H. B. Dewing,* cit.

[185] *Ibidem*

[186] Sołtysiak, A., The plague pandemic and the Slavic expansion in the 6th-8th centuries, cit.

[187] *Ibidem.*

[188] *Ibidem.*

[189] *Ibidem*; Harrison, D., *Plague, Settlement and Structural Change at the Dawn of the Middle Ages,* cit.

[190] Harrison, D., *Plague, Settlement and Structural Change at the Dawn of the Middle Ages,* cit.

[191] Morony, M., *'For Whom Does the Writer Write?'* in Plague and the End of Antiquity : The Pandemic of 541-750, cit.

[192] *Ibidem.*

to Michael the Syrian, those who remained in the cities started to behave like dogs or "attacked one another"[193]; John of Ephesus claim that "a sane person could no longer be found easily"[194]. The same author, in fact, refers to phenomenan of mass hysteria such as the one during which Constantinople's inhabitants started to throw pottery in the street, convinced that this would end the plague. In the city of Amida, which is said to have lost 30.000 people during the pandemic and was in the eight year of famine, people started to behave like sheep and dogs once they received the – false – information that the Persians were coming to conquer their territory. Furthermore, Procopius argues that the cities that were left untouched "were seized with disorder". These reactions were probably triggered by years of wars, famine, and pandemic's outbreaks, which had completely changed society: it is sufficient to say, for instance, that people started to treat "human corpses like the carcasses of animals"[195] and were now "speechless [...], beyond mourning, their hearts grown mute"[196]. People had to accept the multiple burials and the increasing division between the intramural city of the living and the suburban necropolis (literally, the "city of the dead"). Old burial habits were abandoned and multiple burials, as we have seen, appeared all over Europe[197]; people had to accept that they could not have a grave for their loved ones, and this was probably a shock as we can sense from the episode of the man carrying a tablet with his name just to let his relatives whether he had died. Yet, we can only imagine the effect that smelling the bodies piled up in the towers had on the inhabitants of Constantinople, as narrated by Procopius:

> "And when it came about that all the tombs which had existed previously were filled with the dead, then they dug up all the places about the city one after the other, laid the dead there, each one as he could, and departed; but later on those who were making these trenches, no longer able to keep up with the number of the dying, mounted the towers of the fortifications in Sycae, and tearing off the roofs threw the bodies there in complete disorder; and they piled them up just as each one happened to fall, and filled practically all the towers with corpses, and then covered them again with their roofs. As a result of this an evil stench pervaded the city and distressed the inhabitants still more, and especially whenever the wind blew fresh from that quarter."[198]

Burying the dead became the only occupation in Constantinople[199], thus "no one dared to pay any attention to public buildings; there were no public lights in any city, nor any entertainment for the citizens. The theatres, hippodromes, and circuses [...] were all discontinued"[200]. It is not surprising, therefore, that in 556,

---

[193] Micheal the Syrian, in Rosen, W., *Justinian's Flea, Plague, Empire, and the Birth of Europe*, cit.

[194] Incerti auctoris, 2:107; CZ, 112 in Morony, M., *'For Whom Does the Writer Write?*, cit.

[195] Morony, M., 'For Whom Does the Writer Write?, cit.

[196] *Ibidem.*

[197] Hays, J., *Historians and Epidemics*, in Plague and the End of Antiquity: The Pandemic of 541-750, cit.

[198] Procopius, History of the Wars, Book II.22, cit.

[199] Rosen, W., *Justinian's Flea, Plague, Empire, and the Birth of Europe,* cit.

[200] Procopius, The secret History, 26, in Sołtysiak, A., The plague pandemic and the Slavic expansion in the 6th-8th centuries, cit.

after a second outbreak of the epidemic, a three-months-long shortage of bread was followed by riots. Indeed, as Jo Hays clearly points out in *Historians and Epidemics,* pandemics pose a challenge to authorities[201]: for instance, in the seventeenth's century Italian cities this included resistance to quarantine rules, isolation in pest houses, and sanitary cordons, as well as briberies and corruption[202]. On the other hand, pandemics are also instrumental to strengthening of power[203]. This is due to the fact that crises undermine the foundations of power. The aforementioned riots in Constantinople, indeed, were contemporaneous with the worst crisis within the ruling class of the Goths in Spain and many even argue that the effects of the pandemic could have posed the basis for the collapse of the Visigoths' rule in Spain.[204] Moreover, in 560, when it was rumoured that Justinian was dead, there happened to be "bread riots" again and criticism against the Emperor began to spread.

Finally, as we are dealing with a pandemic, we must also mention the relationship between men and the disease, as well as the advancements made in medical science. According to William Rosen, the sixth century medicine was not so different from the one of the beginning of the nineteenth century in treating plague[205]. Galen (129-201) had claimed that the outbreaks of the epidemics were caused by malignant air, the *miasma,* and this would remain the official explanation throughout the Middle Ages[206]; however, others also insisted on astrological explanations[207]. Moreover, according to Stathakopoulos, the presence of talismans suggest that they could have been used against this mysterious disease, as well as magic spells. In fact, little was known about how to deal with the plague, although, for instance, sanitation tactics such as the use of quarantine were widespread[208]. However, as the episode of Bishop Nicholas of Sion (? – 564) demonstrates, it was difficult to put them in practice: when he tried to prevent the farmers to enter his city, he was misunderstood and accused of trying to starve the citizens.[209] In the 7th century, Anastasius of Sinai argued that flight to a location with "healthier airs" could save from the plague[210] (an evident reference to Galen): in premodern times, in fact, when there used to be famines and shortages, people went to the cities, while, on the contrary, when there was an epidemic outbreak, they fled from it[211] . When the plague arrived in Myra, for instance, the farmers refused to bring their products there, fearing the contagion. We can

---

[201] Hays, J., *Historians and Epidemics* in Plague and the End of Antiquity : The Pandemic of 541-750, cit.

[202] *Ibidem:* Corruption was aimed, for example, at ensuring that loved ones were not buried in mass graves.

[203] *Ibidem.*

[204] Kulikowski, M., Plague in Spanish Late Antiquity, cit.

[205] Rosen, W., *Justinian's Flea,* cit.

[206] Stathakopoulos, D., *Crime and Punishment,* in Plague and the End of Antiquity, cit.

[207] *Ibidem.*

[208] Rosen, W., *Justinian's Flea,* cit.

[209] *Ibidem.*

[210] Stathakopoulos, D., *Crime and Punishment,* in Plague and the End of Antiquity, cit.

[211] *Ibidem.*

argue, therefore, that they cannot be properly defined as "measures of quarantine", but rather actions motivated by the observation that plague spread through contact. We have to wait until the second plague pandemic for the modern concept of quarantine[212].

On the other hand, healthcare generally benefitted from the plague, in the sense that new funds were given to medicine: for instance, we know that by 542 the status and the salaries of the doctors had increased[213]. Moreover, even if the reform of the healthcare (the transfer of the civic physicians to the ecclesiastical hospital) was actually implemented ten years before the plague[214], the disease certainly gave an incentive to create new facilities of this kind. However, there is no agreement on the "birth of the clinic" at this time: indeed, John Atkinson claims that the plague invented the "clinical gaze" but did not benefit scientific medicine. On the other hand, Professor Timothy Miller claims that hospitals, in the age of Justinian, provided free specialized care to the sick[215] and Atkinson too agree on the point for which the plague probably incentivised the Emperor to develop health-care facilities[216]: later, hospitals would also become medical schools. However, at the time of Justinian, according to the interpretation of Atkinson of Procopius's word, the *xenones* were at the same time "hostels, poorhouses and hospices/hospitals"[217]: those who could have afforded it were for in their own houses[218].

In conclusion, we cannot be sure that the plague gave the impetus to the spread and development of – what would become – hospitals, although it is somehow revealing that the spread of the hospitals in Europe followed the outbreak of the plague: among the first hospitals we can include the one founded in Montecassino (Italy) by Benedict of Nursia around 540; in France the first one was started in Lyon in 542 and in Paris in 660[219]; in Spain the Catholic Bishop Masona established the first Spanish hospital in Merida[220].

1.1.5 The mystery of the end and of the origin of the Justinian plague

Three years after the first outbreak of the plague, on March 23, Emperor Justinian declared the end of plague[221]. Actually, the epidemic would come back in Constantinople fourteen years later, in 558, and then again in 573/4[222]. Between 541 and 750, eighteen outbreaks were recorded throughout Europe: Evagrius indicates

[212] Quarantena, in Treccani.it – Vocabolario Treccani on line, Istituto dell'Enciclopedia Italiana.

[213] Atkinson, J., *The Plague of 542: not the birth of the clinic*, cit.

[214] *Ibidem.*

[215] Miller, T., *The Birth of the Hospital in the Byzantine Empire,* Johns Hopkins University Press, 1997.

[216] Atkinson, J., The Plague of 542: not the birth of the clinic, cit.

[217] *Ibidem.*

[218] *Ibidem.*

[219] *Hospital,* Encyclopaedia Britannica Online.

[220] Risse, Guenter B., *Mending Bodies, Saving Souls: A History of Hospitals*. Oxford University Press, 1999.

[221] Stathakopoulos, D., *Crime and Punishment,* in Plague and the End of Antiquity, cit.

[222] *Ibidem.*

a periodicity of fifteen years[223], which is indeed more or less the time between an outbreak and another[224]. Then, around 750, the plague mysteriously disappeared[225]. There is no broad agreement on its demise, and it seems that the same hypotheses made for the end of the Black Death cannot be applied to the Justinian plague: neither were quarantine measures, as seen before, widely applicated, nor does another species of mice appear to have replace the *Rattus Rattus* (in the case of the second plague Pandemic, it was the *Rattus norvegicus*)[226]. At the same time, there is no record of any use of rat poisons, no change in human diet and no change in construction techniques[227]: this makes Sallares argue that a possible explanation could be the herd immunity among rodents[228]; on the other hand, McCormick claims that a possible explanation could be the diminished virulence[229]. Moreover, he also introduces the hypothesis that the ancestor of the bacteria *Yersinia Pestis,* could provide immunity to it[230]; "*yersinia pseudotuberculosis,* [...], *yersinia enterocolitica, yersinia pseudotuberculosis* and *yersinia. enterocolitica*"[231], indeed, are naturally overcome by healthy people and cause gastrointestinal disorders. New discoveries seem to confirm that in regions in which gnawers and pigs are affected by the other *yersinia* disease are more resistant to *yersinia pestis* than those in which *Y. enterocolitica or Y. pseudotuberculosis* have not been detected [232]. Therefore, according to McCormick, research should look for presence of these non-mortal forms of *yersinia* infections in territories from which the plague disappeared in the eighth century. Nonetheless, both McCormick and Sallares agree on the necessity of further research to explain the vanishing of the Justinian plague.

Even more mysterious is the origin of the first plague pandemic. We know that it appeared in Pelasium in 541, but nothing more. However, new studies are revealing new evidence and are helping historians and scientists solving one of the most mysterious issues of the medical history.

According to Evagrius, "*It took its rise from Aethiopia, as is now reported, and made a circuit of the whole world in succession*"[233]; Michael the Syrian, reporting John of Ephesus's account, adds something more, arguing that "*It began at first among the peoples inland of the countries of the south-east of India, that is, of Kush, Himyarites, and others; then to the regions of the West*". However, in order to understand more about the spread of the pandemic, we have to analyse why it could have arrived in the port of Pelusium. Tsiamis, Poulakou-Rebelakou and Petridou, indeed, argue that *Yersinia Pestis* could actually have arrived first in the port of Clys-

[223] Evagrius, *Ecclesiastic History*, IV, 29, cit.

[224] Stathakopoulos, D., *Crime and Punishment,* in Plague and the End of Antiquity, cit.

[225] Sallares, *Ecology, Evolution, and Epidemiology of Plague,* cit.

[226] *Ibidem.*

[227] *Ibidem.*

[228] *Ibidem.*

[229] McCormick, *Toward a Molecular History of the Justinianic Pandemic,* cit.

[230] *Ibidem.*

[231] *Ibidem.*

[232] *Ibidem.*

[233] Evagrius, *Ecclesiastic History,* IV, 29, cit.

ma, which at the time had a close trade relation with Pelusium[234]. However, it is problematic to explain why it only arrived later in Egypt's main port, i.e. Alexandria (and not, for instance, contemporaneously)[235]. It has been suggested that the plague may have arrived, on the other hand, via the Silk Road[236], although this does not explain why we have only later accounts of its outbreaks in Persia or Mesopotamia[237]. Evidence, therefore, seems to indicate that it could have travelled along the *maritime* Silk Road, which connected India to the Mediterranean[238]. However, before going on, it must be said that the two superpowers of the time – the Byzantine and the Persian empires – were engaged in a clash for control of the Red Sea; Byzantium exerted influence over the Axumite kingdom, while the Persians on the Himyarites, located on the opposite shore of the sea, in modern Yemen[239]. Since the Himyarites had no complete control over the sea, the Axumite were able to trade as far as Ceylon, where Byzantine coins have been found[240].

Until a few years ago, historians claimed that there was no need to investigate as far as India: they argued that the plague originated in Central Africa and spread from the Horn of Africa to Yemen and then to the Mediterranean[241]. In fact, it was believed that the Great Lakes regions was the most ancient plague reservoir, and that the close contacts between the Axumite kingdom and Byzantium had brought the plague out of Africa[242]. This assertion was based on the studies made by the Belgian medical officer René Devignat, who claimed that the pathogen was carried along the trade routes to Africa before 541 and then spread back from the Great Lakes region[243]. However, new studies from 2013 onwards have confirmed an Asian origin[244]. In particular, one such study, published in 2018, the result of

---

[234] Tsiamis, C., Poulakou-Rebelakou, E., Petridou, E., *The Red Sea and the Port of Clysma. A Possible Gate of Justinian's Plague,* Gesnerus 66,2, 209–217, 2009.

[235] As the author states, the port of Pelusium was "overshadowed by Alexandria".

[236] Wagner, D.M., et al., *Yersinia pestis and the Plague of Justinian 541–543 AD: a genomic analysis,* Lancet Infect Disease, 14: 319-26, 2014.

[237] Shahraki, A.H., Carniel, E., Mostafavi, E., *Plague in Iran: its history and current status*, Epidemiology and Health, 38, 2016.

[238] Boulnois, L., *La Via della Seta,* Bompiani, Milano, 2005.

[239] Tsiamis, C., Poulakou-Rebelakou, E., Androutsos, G., *The role of the Egyptian Sea and Land routes in the Justinian plague: the case of Pelusium,* Medicine and healing in the ancient Mediterranean world, edited by Demetrios Michedelis, Oxbow Books, 2014.

[240] Tsiamis, C., Poulakou-Rebelakou, E., Petridou, E., *The Red Sea and the Port of Clysma. A Possible Gate of Justinian's Plague,* cit.

[241] Keller, M., Spyrou, M.A., et al., *Ancient Yersinia pestis genomes from across Western Europe reveal early diversification during the First Pandemic (541–750),* cit.; Sussman G., *Scientists Doing History: Central Africa and the Origins of the First Plague Pandemic,* Journal of World History, 26, 2, 325-354, 2015

[242] Sussman, G., *Scientists Doing History: Central Africa and the Origins of the First Plague Pandemic,* cit.

[243] *Ibidem.*

[244] Harbeck M, Seifert L, Hanasch S, Wagner DM, Birdsell D, et al., *Yersinia pestis DNA from Skeletal Remains from the 6th Century AD Reveals Insights into Justinianic Plague,* cit.; Damgaard PB, Marchi N, Rasmussen S, et al. *137 ancient human genomes from across the Eurasian steppes*, cit; Eroshenko G., Nosov N., Krasnov Y., Oglodin Y., Kukleva L., Guseva N., et al., *Yersinia pestis strains of ancient phylogenetic branch 0.ANT are widely spread in the highmountain plague foci of Kyrgyzstan*,cit.

collaboration between scientists from around the world, came to suggest that the Huns were responsible of the spread of the Justinian plague. Indeed, analysis carried out on a "Hun from the Thian Shan mountains dating approximately CE 180"[245] revealed the genome of a strain of *Yersinia Pestis* which "branches off from the most basal one"[246] . Interestingly, the most basal one has been found in Xinjiang, very close to the Mountains of Tian Shan[247]. A 2017 study had previously concluded that the same strains later found in the 180 CE's Hun, "are ones of the most closely related to the *Y.pestis* strain responsible for the Justinian plague"[248] Further analysis concluded that the strain found in the individual dating to CE 180 had "full bubonic capability and flea transmissibility"[249]. The authors of the 2018 study thus concluded that the bubonic plague was took to the West either through the Silk Road or thanks to the westward movement of the Huns [250]. However, if this is true, it does not explain the problem mentioned above, that is to say why we don't have accounts of previous outbreaks of the Justinian plague in the countries along the Silk Road or in the southern Steppe. Huns, indeed, arrived in Crimea by 395[251], but no outbreak of the plague was recorded north of Constantinople before 542. In fact, if the plague had broken out before 542 in territories close to the Huns settlements, it would have previously hit in the Empire's capital. Again, therefore, the evidence seems to suggest that we need to focus on the *maritime* Silk Road. Interestingly, since in 525, the Kingdom of Axum conquered the lands of the Hymiarites[252] from then onwards there were no more obstacles to trade between the Indian Ocean and Axum (and therefore, to Byzantium). Meanwhile, in the Indian subcontinent, the Hephthalites Huns gained control of the "most important section of the Silk Road", the one extending from the Tarim Basin to the port of Bharuch near the Indian ocean[253]. The origin of the Hephthalites is still debated: according to Procopius, the Hephthalites were "of the stock of the Huns in fact as well as in name; however, they do not mingle with any of the Huns known to us"[254]; his contemporary, Agaphius of Myrinae states that "the Hephthalites are a Hunnish people"[255]. Interestingly, in the Chinese "History

[245] Damgaard PB, Marchi N, Rasmussen S, et al. 137 ancient human genomes from across the Eurasian steppes, cit.

[246] *Ibidem.*

[247] *Ibidem.*

[248] Eroshenko GA, Nosov NY, Krasnov YM, Oglodin YG, Kukleva LM, Guseva NP, et al., *Yersinia pestis strains of ancient phylogenetic branch 0.ANT are widely spread in the highmountain plague foci of Kyrgyzstan,* cit.

[249] Damgaard PB, Marchi N, Rasmussen S, et al. *137 ancient human genomes from across the Eurasian steppes,* cit.

[250] *Ibidem*

[251] Heather, P., *La caduta dell'Impero romano: una nuova storia*, Garzanti, Milano 2006.

[252] "Himyar", Encyclopaedia Britannica online.

[253] Harmatta, J., *Annexation of the Hephthalite vassal kingdoms by the western Turks* in History of Humanity: From the seventh century B.C. to the seventh century A.D, edited by Joachim Herrmann and Erik Zurker, UNESCO, 1996.

[254] Procopius 1914 - I, 12-15, in Kurbanov, A., The Hephthalites: archaeological and historical analysis, Berlin Freie Universität Berlin, 2010.

[255] Гафуров 1972, 206 in Kurbanov, A., *The Hephthalites: archaeological and historical analysis*, cit.

of the northern kingdoms", dated to 644, it is reported that "they originated from the north of the Chinese frontier and came down south from the Jinshan mountain"[256]. Moreover, they practiced the deformation of the head, a custom typical of the Huns[257].

The Hephthalites established their rule in a period of intense instability, testified by the decline of irrigation, urban centres and pottery[258]: archaeologist found several deserted towns dating to the 4th-5th century and many speak about a social decline typical of the 4th-6th centuries[259]. Interestingly, several researchers have linked this sudden abandonment to the spread of a contagious disease, perhaps the plague[260]. Moreover, Procopius reports about collective burials of the Hephthalites[261]. We have already seen that this is typical of contagious diseases; according to Ilyasov, however, the Chinese chronicle "Liangshu" reports that Hephthalites were buried in coffins, therefore the collective burials were not their typical burial custom. Unfortunately, we cannot neither verify the statement of the Byzantine historian, nor hypothesise that he was not referring to ceremonial mounds, for instance. However, it is rather interesting that a nomadic population, contemporary with the first outbreak of the Justinian plague, bordererd both an important seaport of the maritime Silk Road and one of the natural foci of the plague, the Tarim Basin, from where we now know that the first plague pandemic originated. The plague may have moved from the mountains of Tian Shan in the first decades of the 6th century, arrived in India and then reached the shores of the Red Sea by maritime trade. Importantly, northwest India is still nowadays one of the natural foci of the plague among roditors, along with Central Asia[262]. Alternatively, it has been suggested that the plague cfirst reached Ethiopia and from there it arrived in Egyptt[263]. However, importantly, according to Robert Sallares:

> "To produce human epidemics of bubonic plague, a triad of organisms is needed: plague itself, the black rat (Rattus rattus), and the rat flea Xenopsylla cheopis. An Asian origin is still possible for plague itself, while the rat came from southern Asia. However, the flea X. cheopis is thought to be native to northeastern Africa. Consequently, it is possible, and indeed quite likely, that Egypt or East Africa is where these three organisms came together for the first time. It is not surprising that the Justinianic Plague began in Egypt."[264]

---

[256] Бичурин Н.Я. (Иакинф) 1950, 205; Vaissière 2003, 125 in Kurbanov, A., *The Hephthalites: archaeological and historical analysis*, cit.

[257] Kurbanov, A., *The Hephthalites: archaeological and historical analysis*, cit.

[258] *Ibidem*.

[259] *Ibidem*.

[260] Schamiloglu, U., *The Plague in the Time of Justinian and Central Eurasian History: An Agenda for Research* in O. Karatay, & I. Zimonyi (Eds.), Central Eurasia in the Middle Ages: Studies in Honour of Peter B. Golden, Turcologica, 104, Harrassowitz Verlag.

[261] Ilyasov 2003, 143., in Kurbanov, A., *The Hephthalites: archaeological and historical analysis*, cit.

[262] Bazanova, L.P. et al., *Natural foci of plague in India: epizootological characteristics*, Meditsinskaia Parazitologiia i Parazitarnye Bolezni, July 2009.

[263] Gebre Selassie, Y., *Plague as a Possible Factor for the Decline and Collapse of the Aksumite Empire: a New Interpretation*, Northeast African Journal of Social Sciences and Humanities, 1, 2011.

[264] Sallares, R., *Ecology, Evolution and Epidemiology of Plague*, cit.

Therefore, according to Robert Sallares, the so called "spill-over" may have taken place in Egypt: from being a rodent's disease, the plague turned into a human disease. Then, from Pelusium, the plague throughout the Mediterranean and beyond. This would also explain why the plague seems to have arrived in the Sasanid Empire from the West (Egypt) rather than from the East (India/China). However, according to the new interpretation mentioned above, the plague might have also spread "human to human" thanks to the actions of the human flea such as *P.irritans*: therefore, in order to verify the hypothesis of Sallares, we should know the geographic distribution *at the time* of the *Xenopsylla cheopis* and of *Pulex irritans*.

In conclusion, for now and until new research and discoveries are made, the origin of the first plague pandemic remains a mystery. We only know that it originated in Central Asia, but we don't know how it arrived in Europe, whether due to the wandering of the Hephthalites from the Tamir Basin to the Indian Ocean or via the trade routes of the time. Great changes were taking place at the time: the Huns, were indeed moving from the inner Asia and provoking a *domino effect* which would have led the Germanic populations into the Roman Empire, triggering the end of an era.

1.2 Migrants at the turn of the Middle Ages: The Barbarian invasions

There is no agreement on the definition of the time limits of the *"Migration period"*: according to several sources, in fact, it could be a synonym of the Early Medieval period[265], while for others it is limited to the 6th century[266]. However, since we are focusing on the issue with a *longue durée* approach, we argue that the first definition should be adopted: in our view, indeed, the Migration Period must be seen as an era of intense social changes and upheavals of people who, for different reasons, migrated to the territory of the Roman Empire. Since the eighteenth century, the so called "barbarian invasions" have been regarded as one the main causes of the collapse of the Roman Empire[267]; this explanation is still accepted today, although many other historical trends are underlined. These include, for instance, economic and demographic reasons[268]. Nonetheless, we will not focus on these explanations, as we will mainly analyse the process of departure and settlement of the Germanic tribes in Western Europe, stressing how, rather than "invasions", this process should be seen as a centuries-long mass migration[269] that pushed northern tribes "down upon earlier emigrants to the south"[270] and was triggered by several causes.

---

265 *"Migration period"*. Encyclopedia Britannica.

266 Drake, B., Lee, *Changes in North Atlantic Oscillation drove Population Migrations and the Collapse of the Western Roman Empire*, Sci Rep, 7, 1227, 2017.

267 Middleton, G.D., *Understanding Collapse: Ancient History and Modern Myths*, Cambridge University Press, 2018.

268 Heather, P., *La caduta dell'Impero Romano: una nuova storia*, cit.

269 Vogt, J., *The decline of Rome,* Richard Thornton Books, London, 1967.

270 Goffart, W.A., *Barbarians and Romans, A.D. 418-584: The Techniques of Accommodation*, Princeton University Press, Princeton 2020.

1.2.1 The migratory movements

A first movement took place around 1000 BC: at the time, the Germanic Peoples of Scandinavia, Denmark, and Northern Germany started to move southwards, displacing the Celts[271]; then, between 600 and 300 BC, a second group moved from Scandinavia to the southern shores of the Baltic, in present-day Poland. J.B. Bury designates the former as "west Germans" and the latter as "east Germans". In his *De origine et situ Germanorum,* Tacitus enumerates the different Germanic tribes' names which, however, will disappear in the following centuries[272]. Tacitus's importance, however, lies in the fact that he is the only "surviving ethnographic treatise from the ancient world concerned only with the northern barbarians"[273]. Moreover, according to J.B. Bury, in the century between Cesar's account (*De Bello Gallico,* I BC) and Tacitus's one (II CE), the Western Germans changed their habits, and from a "pastoral" and "migratory" population they became an "agricultural" population[274]. On the other hand, the eastern Germans, not having encountered theg Romans, did not abandon their pastoral way of living.

These two distinct processes were not the only ones that took place in *Barbaricum,* the name Romans gave to the lands beyond the *limes,* the border. The population growth, which had previously displaced the Germanic peoples from their original homeland, was now reducing the distance between the tribes[275]. Once separated by rings of forest, the different communities were becoming closer than ever: hence, the birth of the political confederations such as the *Alemanni* ("all men"[276]), a composition of the Suevian tribes in the Upper Rhine, the Franks (the "free" ones), a "loose conglomerate" in the lower Rhine[277], the Saxons, in modern-day Schleswig-Holstein, and the Thuringians, the ancient Hermunduri[278]. This federation process also saw the emergence of a warrior aristocracy, which will evolve into the later feudal nobility[279]. Various researches argue that this process was also favoured by the proximity of the Empire, whose organization and status provided an example to be followed[280].

We will focus on this claim later. In the meantime, the Eastern Germans were moving slowly throughout Eastern Europe, probably down to the Black Sea, where we find a first group of them, the Goths, in 214 CE [281]. Interestingly, they almost certainly followed the same route that the Varagians/Swedish Vikings will follow later. Probably because of the distinction between newcomers and their predecessors[282], the

[271] Bury, J.B., *The Invasion of Europe by the Barbarians,* Jovian Press, ebook, 2017.

[272] *Ibidem.*

[273] James, E., *Europe's Barbarians AD 200-600,* Taylor and Francis, ebook, 2014.

[274] Bury, J.B., *The Invasion of Europe by the Barbarians,* cit.

[275] *Ibidem.*

[276] Meeson Beck, F.G., *Alemanni,* Encyclopaedia Britannica, Volume 1, 1911.

[277] Bury, J.B., *The Invasion of Europe by the Barbarians,* cit.

[278] *Ibidem.*

[279] Ruffolo, G., *Quando l'Italia era una superpotenza,* Einaudi, Torino, 2004.

[280] Halsall, G., *Barbarian Migrations and the Roman West, 376-568,* Cambridge University Press, Cambridge 2007.

[281] Bury, J.B., *The Invasion of Europe by the Barbarians,* cit.

[282] *Ibidem.*

people, once united within a single confederation[283] began to perceive themselves as two divided groups, the Visigoths and the Ostrogoths[284]. We will briefly introduce the migration of the Goths so to describe several trends that were typical of the period.

This mass migratory movement of the Goths from Scandinavia did not stop on the shores of the Black Sea; from there, in fact, the Goths became the first Germanic tribe to settle within the Empire. After moving as far as to the Aegean shores[285], the Goths settled in Dacia in 256 CE, where they were granted the status of *foederati* in 332, i.e. they accepted to protect the Empire and to provide soldiers to the Imperial Army in exchange for subsides and lands; Dacia became nominally part of Constantine's Empire, although occupied by the Visigoths[286]. In fact, this process of federation became the norm for the Germanic tribes that pushed to the borders of the Empire[287]. Interestingly, the movement of the Goths had previously been at the origin of the first barbarian wars and the prelude of the great migrations of the Germanic tribes: the Marcomannic Wars fought by Marcus Aurelius. Both Bury[288] and Goffart[289] agree that the Goths' movement displaced the tribes bordering the Empire, who, due to the lack of places to resettle, were forced to enter the lands of the Empire. Importantly, this process of pushing populations into the Empire will be a constant of the following centuries. However, the view for which the migrations of the Germanic tribes should be only related to the push of other peoples, has been recently challenged by Goffart[290]: in his vision, we should also look at the relationship between both sides of the borders of the *limes* and focus on the invitation of settlement made by several roman authorities throughout the Migration Period. Therefore, in order to better analyse these political changes, we will briefly describe the migratory movements that changed the history of Europe between the III and the 6th-7th centuries.

1.2.1.1 The Gothic Crisis: Italy, Gaul, and the Iberian Peninsula

What is commonly known as "Gothic crisis" describes the settlement and the revolt of the abovementioned Goths within the border of the Roman Empire: pushed by the arrival of the Huns from the eastern steppe[291] the Visigoths begged to be allowed to enter the Roman borders in safety. Curiously, Halsall describes them as "refugees",[292] and, effectively, the given description resembles familiar images. According to Amminius Marcellinus:

---

[283] Southern, P., *The Roman Empire: from Severus to Constantine*, Londra & New York, Routledge, 2001.

[284] As Jordanes states in the *Getica*, "Ostrogoths and Visigoths, were united" (Jordanes, *The origin and the deeds of the Goths*, translated by Charles C. Mierow).

[285] Halsall, G., *Barbarian Migrations and the Roman West, 376–568*, cit.

[286] Bury, J.B., *The Invasion of Europe by the Barbarians*, cit.

[287] *Ibidem*.

[288] *Ibidem*.

[289] Goffart, W.A., *Barbarians and Romans, A.D. 418-584: The Techniques of Accommodation*, cit.

[290] *Ibidem*.

[291] Halsall, G., *Barbarian Migrations and the Roman West, 376–568*, cit.

[292] *Ibidem*.

"In the meantime, a report spread far and wide through the nations of the Goths, that a race of men, hitherto unknown, had suddenly descended like a whirlwind from the lofty mountains, as if they had risen from some secret recess of the earth [...]. And then the greater part of the population resolved to flee and to seek a home remote from all knowledge of the new barbarians; and after long deliberation as to where to fix their abode, they resolved that a retreat into Thrace was the most suitable for these two reasons: first of all, because it is a district most fertile in grass; and secondly, because, owing to the great breadth of the Danube, it is wholly separated from the districts exposed to the impending attacks of the invaders. Accordingly, under the command of their leader Alavivus, they occupied the banks of the Danube, and sent ambassadors to the emperor Valens, humbly entreating to be received by him as his subjects. They promised to live quietly, and to furnish a body of auxiliary troops if necessary."[293]

The response of Rome was to accept the asylum seekers and:

"as soon as they had obtained permission of the emperor to cross the Danube and to cultivate some districts in Thrace, they poured across the stream day and night, without ceasing, embarking in troops on board ships and rafts and on canoes made of the hollow trunks of trees..."[294]

However, the Visigoths would have later rebelled because of the lack of food and once they were moved southwards, another German tribe, the Greuthungi, crossed the Danube and gathered troops of Roman army deserters, miners, and slaves, and started to raid the territory up to Constantinople[295] . The Empire was then forced to sign a treaty and the Goths were settled – and divided – among the Balkans and they gave up their ambition to self-rule themselves. However, these people, warriors, but also women and children, revolted again in 395, under the lead of Alaric, who, in an attempt to force the Roman into a new peace[296] first came down to Athens and then, once he failed to obtain new settlements in Moesia and Macedonia, led the Visigoths up to Rome, trying to reach an agreement with the General Stilicho. Even after the famous sack of Rome in 410, Alaric continued to demand land in Noricum and Dalmatia, as well as the title of *magister militum*[297], i.e. "General". This clearly shows that Barbarians did not want to destroy the Empire, but rather to be accepted into it and obtain the same status of its citizens. Alaric even tried to do what Odoacer later did: ask the Senate to replace the Emperor with someone keener to the Goths. Then, having achieved nothing, the Visigoths moved to Gaul and then to Spain; meanwhile, Alaric's brother, Ataulf, had married Galla Placidia, daughter of the Emperor Theodosius. According to James, indeed, Ataulf wanted to support the Roman Empire through the strength of the Goths[298]. However, they were granted a settlement in the Garonne valley in Gaul

[293] Marcellinus Ammianus in Robinson, J.H., *Readings in European History,* Boston, New York, Ginn and co., 35-39, 1905.

[294] *Ibidem.*

[295] James E., Europe's Barbarians AD 200-600, cit.

[296] *Ibidem.*

[297] *Ibidem.*

[298] *Ibidem.*

only in 418, after having been recognized as *foederati* in exchange for their commitment to fight against the Alans and the Vandals in *Hispania*. This would become the starting place of their kingdom over Southern Gaul and the great part of the Iberian Peninsula; the Goths managed to establish good relations with the Roman aristocracy in the area and protected it from other barbarian incursions until the arrival of the Franks, who would put an end to their kingdom in modern-day France in 507[299]. The Visigoths in the Iberian Peninsula, on the other hand, were defeated by the Arabs in 711[300].

Before that, however, the Visigoths in Spain, had defeated the Vandals, the Alans and the Suebi who had left the lower Danube region to cross the border around 400. Although Procopius claims that the three tribes migrated due to the famine, Merrills argue that their motive was the same as the Visigoths' movement: the arrival of the Huns[301]. MacDowwall compares the Vandals, the Alans and the Suebi to the modern migrants crossing the Mediterranean, as they were so desperate and impoverished to risk a migration in the middle of the winter[302]. After encountering a Frankish army sent by the Empire, this conglomerate of people moved southwards to Spain, where they settled. It is debated whether they received the status of *foederati* once arrived in *Hispania*[303]. However, they only remained in peace less than a decade, and were then forced to move in North Africa: Between 50.000 and 80.000 people, including men, both young and old, women, children, and slaves[304] left the southernmost point of Spain in 429. An entire population which had left the forests of Central Europe only twenty-five years before, was now entering the sandy costs of the Roman province of *Mauretania*.

1.2.1.2 From Gaul to France: The Franks, the Huns and the Franks again

As in the other parts of the Empire[305], the third century saw a general crisis in France, testified by the build of walls far from the borders and the great finds of gold hoards, which seem to reveal fear and displacement of people[306]. In this period the inhabitants of Gaul felt abandoned by the Empire, that was concentrating his forces on the defence the East and Italy; moreover, it was in this very period that the first incursions of the Franks took place, such as the one that started in Lower Rhine and ended in Africa, more or less like the subsequent one of the Vandals[307]. However, with the appointment of Maximianus as co-ruler of Diocletian in the West, the Empire was able to restore order in this region, thanks to the fleet used to patrol the English

[299] *Ibidem*.

[300] Halsall, G., *Barbarian Migrations and the Roman West, 376–568*, cit.

[301] Merrills A, *Vandals, Romans and Berbers: New Perspectives on Late Antique North* Africa, Taylor and Francis, ebook, 2017.

[302] MacDowall, S., *The Franks*, Pen and Sword, Barnsley, UK, 2018, 2018.

[303] Sivan, H., *On Foederati, Hospitalitas, and the Settlement of the Goths in A.D. 418*, The American Journal of Philology, 108, 4, 759-772, The Johns Hopkins University Press, 1987.

[304] Merrills A, *Vandals, Romans and Berbers: New Perspectives on Late Antique North* Africa, cit.

[305] Halsall, G., *Barbarian Migrations and the Roman West, 376–568*, Cambridge University Press, cit.

[306] MacDowall, S., *The Franks*, Pen and Sword; cit.

[307] *Ibidem*.

Channel and the Rhine. In the fourth century, the Franks were based in the Lower Rhine, while the Alamanni in the east bank of the Upper Rhine; these tribes alternatively joined the Romans against other tribes or fought the Empire. However, in the second half of the fifth century, a part of the Franks (the *Salii*) became *foederati*, providing military help against the incursions of the other Germanic tribes (as they did against the abovementioned Alans and Vandals)[308]. The Franks would then be divided between those who invaded Gaul with the forces of Attila and those who defended the province under the lead of General Aetius[309]; Jordanes, indeed, writes in his *Getica*:

> "On the side of the Romans stood the Patrician Aëtius, on whom at that time the whole Empire of the West depended; a man of such wisdom that he had assembled warriors from everywhere to meet them on equal terms. Now these were his auxiliaries: Franks, Sarmatians, Armoricians, Liticians, Burgundians, Saxons, Riparians, Olibriones and some other Celtic or German tribes.[310]"

The Huns, which were now threatening Gaul, were probably born from a part of a Mongol tribe, the Xiong-Nu, in the Far East, from which they moved during the first centuries of the first millennium. Wandering throughout Eurasia this tribe became a conglomerate of peoples[311] and finally reached Europe around 370 CE, when they were responsible of the domino-effect which pushed Germanic tribes into the Roman Empire[312]. Indeed, while some peoples decided to submit themselves to the newcomers, other decided to flee: some Alans, for instance, were absorbed in the Huns, while others fled westwards[313]. Moreover, once arrived in the Danubian region around 380, the Huns pushed other Germanic tribes into the Empire; for instance, this was the case of the Suebic Juthungi, who moved into Gaul but were repelled by the Empire and were finally defeated by the Alamanni[314]. Interestingly, when some Goths left the Huns for Italy in 405, a Hun leader, Uldin, allied himself with the Roman General Stilicho to punish this secessionist attempt: according to the historian Kim, this willing to cooperate with both the Western Roman Empire and with the Eastern one must be related to the process of consolidation in the Hun confederation, as well as to the need to concentrate on the expansion into Western Germany. Remarkably, it was in the same period, during 405/6 that the Alans, the Suebi and the Vandals left *Barbaricum* for Gaul. Moreover, General Aetius regained control of Gaul thanks to the alliance with the Huns. Aetius also came to an agreement with the use of the Germanic tribes, who became *foederati* and settled in Gaul: Alans and Burgundians helped to suppress

---

[308] *Ibidem.*

[309] *Ibidem.*

[310] Jordanes, *The origin and the deeds of the Goths*, cit.

[311] Brosseder, U. B., *Xiongnu and Huns*, Archaeological Perspectives on a Centuries-Old Debate about Identity and Migration in Empires and Exchanges in Eurasian Late Antiquity Rome, China, Iran, and the Steppe, ca. 250-750, edited by Nicola di Cosmo, Institute for Advanced Study, Princeton, New Jersey, Michael Maas, Rice University, Houston, 2018.

[312] Kim, H.J., *The Huns*, Taylor and Francis, ebook, 2015.

[313] *Ibidem.*

[314] *Ibidem.*

several riots also known known as *bacaudae*. However, this process of settlement was damaging the control of the Empire, although, according to Ravegnani, another solution was not possible, since the Romans were not able to stop the arrivals of these Germanic tribes[315]. Then, after reaching an agreement with the Franks, who had invaded Gaul in 446[316], Aetius managed to defeat the Huns, who, led by their new leader Attila, had invaded modern-day France[317]. nterestingly, decades of migrations had actually made the two armies similar: on both sides there were German tribes, Gallo-romans, Goths (Visigoths in the Empire's army, Ostrogoths in Attila's) and even the Huns (some of whom had collaborated with the Romans in the previous years)[318].

Aetius' victory, however, did not allow the Roman rule to survive in Gaul[319]; the baucaudae did not stop and areas not in control of the Empire began to expand. Eventually, some Britons, fleeing the Saxons in Britannia, came to settle in Ar morica and joined the insurgents. In the following years, Gaul became divided between different warlords, in a situation of increasing social disorder; after the deposition of Romulus Augustulus by Odoacer, Childeric emerged among the Franks, and by the end of his life in the 480s, he had posed the basis of the Frankish kingdom and gained the support of some of the Gallo-Romans[320]. His son Clovis would defeat the last Gallo-Roman warlord and from that point onwards Germanic tribes would rule in Gaul[321].

1.2.1.3 *Britannia:* the Anglo-Saxons' arrival and the flee of the Romans

Bede is widely regarded as the main historical source for the arrival of Jutes, Angles and Saxons in *Britannia*[322]*; in his comment of Book I, chapter 15 of the Ecclesiastic History,* he states that these people were among the strongest of *Barbaricum*[323]. A great debate, however, has taken place in the last decades on the issue of the numbers and the origin of the migrants that arrived in modern-day England[324]: should we consider it a mass migration, or a movement limited to a few small groups? New studies, based on technological analysis, argue that the Anglo-Saxons' one was probably a large-scale immigration[325]. Nonetheless, the beginning of this process is still a mystery: findings from Anglo-Saxon's cemeteries seem to point in the direction of the arrival of these Germanic tribes as *foederati* or at least

[315] Ravegnani, G., *La caduta dell'Impero Romano*, Il Mulino, Bologna 2012.

[316] *Ibidem.*

[317] MacDowall, S., *The Franks*, cit.

[318] *Ibidem.*

[319] *Ibidem.*

[320] Wood, I., *Before and after the Migration to Britain in The Anglo-Saxons From the Migration Period to the Eight Century*, an ethnographic perspective, edited by John Hines, Boydell Press, Woodbridge, 2007.

[321] *Ibidem.*

[322] Harke, H., *Anglo-Saxon Immigration and Ethnogenesis*, Medieval Archaeology, 55, 2011.

[323] *Ibidem.*

[324] Jones, M.U., *Mucking and Early Saxon rural settlement in Essex,* in Archaeology in Essex to AD 1550, Edited by D.G., Buckley, CBA Research Report, 1980

[325] Myres, J. N. L., *The Anglo-Saxon Settlements,* Oxford University Press, 1969

as mercenaries employed by the Romans[326] and this seems to be confirmed by the fact that their initial settlements were close to the Romans' ones[327]. However, it is difficult, at least for the first settlements, to claim that only the three tribes mentioned above moved to *Britannia:* most of the earliest burials, in fact, are not only "Jutish, Anglian or Saxon", but also "Frankish, Alemannic and Thuringian"[328] and even "Scandinavian".[329] It is also difficult to isolate the earliest archaeological evidence of the Saxons from that of the Franks in Continental Europe, for instance; furthermore, while we know enough about the Danes, on the other hand, we know almost nothing of the Jutes and Angles[330]. Equally obscure are the reasons that led these tribes to leave for *Britannia*. The Byzantine historian Zosimus claims that while the Emperor Constantine III – a British – left *Britannia* for Gaul in 407 to face the invasion of Vandals, Alans and Suebi, who were advancing up to the English Channel, Britain saw the arrival of "barbarians from beyond the Rhine"[331] and a Gailic Chronicle adds that the country was "devasted by an incursion of Saxons".[332] Again, we are dealing with a migration of a Germanic tribe in the crucial years 405/6, during which the Huns were moving within *Barbaricum*. Archaeology has confirmed that previously inhabited settlements along the Frisian and North German coasts were abandoned during the fifth century and at the same time, according to the archaeologist Hawkes, we find "evidence for the appearance of cremating peoples, from continental Anglia and Saxony, in parts of Eastern England and in the Midlands during the early years of the century". That is why the archaeologist call these migrants "boat people"[333] rather than raiders: Britain, indeed, had faced maritime raids of the same Germanic people in the 3rd century; however, as the Vikings would do centuries later, these tribes eventually settled there – though for different reasons. Interestingly, according to Todd Morrison, the very name *"Saxons"* is too widespread in Late antiquity to indicate only one tribe: he claims, therefore, that this name should be rather seen as a synonym for "pirate". These pirates from different tribes would have later formed a confederation, known as "Saxons", which would have then started settling in other lands. Again, this would be the same for the vikings. According to one theory, in fact, the very name *"Saxon shore"* – i.e. the system of fortifications on both sides of the English Channel – could derive its name both from the raids of the said population and

---

[326] Wood, I., *Before and after the Migration to Britain* in The Anglo-*Saxons From the Migration Period to the Eight Centurym,* cit.

[327] *Ibidem*

[328] *Ibidem*

[329] Zosimus, vi.5.2 f in Hawkes, S.C., *Anglo-Saxon Kent c 425-725,* in Archaeology in Kent to AD 1550: in memory of Stuart Eborall Rigold, Edited by D.G., Buckley, CBA Research Report, 1982

[330] Hawkes, S.C., *Anglo-Saxon Kent c 425-725,* in Archaeology in Kent to AD 1550: in memory of Stuart Eborall Rigold, cit.

[331] *Ibidem*

[332] Morrison, T., *Britain and the Anglo-Saxons in Late Antiquity,* University of Mexico Digital Repository, 2016.

[333] According to many studies, the name *Vikings* comes from "vik", bay, thus being a way to indicate mariners or pirates. However, it would become a name to indicate an historical period and, even though incorrectly, a population.

from their previous settlement; moreover, the same Saxons could have been settled in northern Britannia and fought against the Picts, laying the foundations for the later mass movement[334]. Indeed, according to Gildas, they were settled in order to "repel the northern nations"[335].

As for the numbers of the migrants, new studies based on the analysis of the DNA claim that between 100.000 and 200.000 migrants may have arrived in Britain during the fifth century: this would imply that the estimated proportion of the newcomers to natives may have been around 10%[336]. According to Harke, this is in line with the numbers of the Alans and Vandals who moved to Spain (80.000) and of the Ostrogoths who left the Balkans for Italy (100.000).

These movements took place contemporary to the departure of the Romans from *Britannia*; in the same year 406, in fact, the Roman troops in Britannia elected their own Emperor, Marcus, who was soon replaced by a town councillor, Gratian, and again by the above-mentioned Constantine[337]. Actually, in 383, the Commander of the Roman legions in Britain, Maximus Magnus Clemens, had induced his troops to proclaim him emperor[338]. This alternance of rulers gives the idea of a general instability, also triggered by the situation in continental Europe – for which the newly elected Constantine almost immediately left Britain – and by the continuous raids of the said Saxons and even of the northern tribes of the Scots and the Picts, as narrated by Marcellinus Ammianus in his chronicles[339]. With the emperor gone, and facing the barbarian threat, the Romano-Britons rebelled and expelled the Roman officials[340]. It is dubious, however, whether the Britons rebelled against Rome, or the Roman Empire left Britain due to the urgency to face the growing Barbarian threats[341]. Generally, the process of settlement of the Anglo-Saxons was accompanied by another process of migration: the one of the Britons, who left the east coast of Britannia, fleeing the arrival of the newcomers[342]. This period of social unrest, indeed, made some of the native Britons move to *Armorica*, thus renaming this land in *Brittany*[343]. This event is also recorded by the contemporary Procopius, who states:

---

[334] Elverhøi, Y. A., *Germanic piracy in Roman Britain*, A Thesis Presented to The Department of Literature, Area Studies and European Languages, University of Oslo, 2010.

[335] Gildas, The Ruin of Britain, cit.

[336] Harke, H., *Anglo-Saxon Immigration and Ethnogenesis*, cit.

[337] Johnson, S., *Later Roman Britain*, Taylor and Francis, ebook, 2014.

[338] Shearman, J.F., *The Celto-Britons of Armorica-Maximus Clemens, Emperor of Britain, Gaul and Spain; The British Legions; Conan of Merladog, His Descendants Kings of Britagne: Continued Intercourse with the Mother – Country*, Armorican Saints and Ecclesiastics; *Their Connexions with Great Britain and Ireland in the Fifth and Sixth* Centuries, The Journal of the Royal Historical and Archaeological Association of Ireland, 5, 49, 597-630, 1882.

[339] *Le Storie di Ammiano Marcellino*, tradotte da Francesco Ambrosoli, Antonio Fontana editore, Milano, 1829.

[340] Johnson, S., *Later Roman Britain*, cit.

[341] Jones, M., *The end of Roman Britain*, Cornell University Press, Ithaca, New York, 1998.

[342] Dark, K.R., *Large-scale population movements into and from Britain south of Hadrian's Wall in the fourth to sixth centuries AD*, Reading Medieval Studies, 29, 31-49, University of Reading, 2003.

[343] "Brittany". Encyclopedia Britannica online.

"The island of Brittia is inhabited by three very numerous nations, each having one king over it. And the names of these nations are Angili, Frissones, and Brittones, the last being named from the island itself. And so great appears to be the population of these nations that every year they emigrate thence in large companies with their women and children and go to the land of the Franks. And the Franks allow them to settle in the part of their land which appears to be more deserted, and by this means they say they are winning over the island. Thus it actually happened that not long ago the king of the Franks, in sending some of his intimates on an embassy to the Emperor Justinian in Byzantium, sent with them some of the Angili, thus seeking to establish his claim that this island was ruled by him. Such then are the facts relating to the island that is called Brittia.[344]"

Other Britons are said to have reached even Galicia, in modern-day Spain; this is proved by the presence of a bishop named Mailoc – a Celtic name – representative of a *Britonensis Ecclesia* at the second council of Braga in 572, as well as other representatives for this diocese, recorded in 633, 646, 653 and 675[345].

Furthermore, the migration of the Anglo-Saxons and the emigration of the Britons were not the only people's movement which took place in Britannia: we have already mentioned, indeed, the simultaneous raids of the Picts and of the Scots, an Irish tribe. According to some studies, these invasions must be seen as earlier stages of settlements, at least for what concerns the Irish tribes[346]. Philip Rance, indeed, recalls the appearance of the *Attaccotti* tribe among those who raided Britain in the 360s, later among those who composed the Roman Army and then in the settlements in Western Britain of the *"aithechthuaata"*[347]. We lack knowledge about the extent of the settlements, as well as about duration of this process or the role these troops had in the defence of the Empire; however, this is perfectly in line with the general process that saw the border tribes welcomed in the Empire as *foederati*, as seen above. On the other hand, despite the previous raids, the Picts generally remained in their territories, and did not settle south of the Adrian's Wall[348].

1.2.1.4 From the Ostrogoths to the Lombards: the end of the Migration Period?

After the sack of Rome in 410 by the Visigoths, Italy had to face the invasion of the Huns in 452[349], and three years later, the Vandals, coming from Carthage[350]. In

[344] The Wars of Procopius, cit., Book VIII, chapter XX.

[345] Tovar, A., *Un obispo con nombre británico y los orígenes de la diócesis de Mondoñedo,* Habis, 3, 155-158, 1972.

[346] Rance, P., *Attacotti, Déisi and Magnus Maximus: The Case for Irish Federates in Late Roman Britain*, Britannia, Society for the Promotion of Roman Studies, 32, 243-270, 2001; Dark, K.R., *Large-scale population movements into and from Britain south of Hadrian's Wall in the fourth to sixth centuries* AD, cit., Rance, P., *Epiphanius of Salamis and the Scotti: New Evidence for Late Roman-Irish Relations, Britannia*, The Society for the Promotion of Roman Studies, 2012.

[347] Rance, P., Attacotti, Déisi and Magnus Maximus: The Case for Irish Federates in Late Roman Britain, cit.

[348] Halsall, G., *Barbarian Migrations and the Roman West, 376–568,* cit.

[349] Kim H.J. *The Huns,* cit.

[350] Merrills A., editor. Vandals, Romans and Berbers: New Perspectives on Late Antique North, cit.

fact, having arrived in Africa from Central Europe, the Vandals returned to the European continent, conquering Sicily, Sardinia, and Corse and raiding Rome for two weeks: then they returned to Africa [351]. Yet, Roman imperial authority endured and so did the social and cultural structure, despite the growing importance of the barbarian influence in the Army: indeed, due to the military power, the Germanic tribes were becoming more and more able to have a say in political decisions[352]. In fact, it was a Germanic mercenary leader, Odoacer, who deposed the last Roman Emperor, with the aim to obtain lands where he could settle with his tribe[353]; he didn't want to overthrow the Empire, and he even adopted a policy of pacification, seeking a peaceful coexistence between the Germanic armies and the Roman moral and political authority[354]. We don't know exactly Odoacer's origin, but he is described by ancient authors indifferently as a Sciriian, a Rogian, a Torcilingian a Herul or a Goth[355]; this gives an idea of the ehtnic fragmentation and confusion triggered by the migrations of that period. For instance, a group of the Heruls, whose Odoacer is described as the king, is said by Procopius to have moved from Central Europe to "Thule" – probably Scandinavia, from which the Heruls probably came from, as revealed by the episode narrated by Procopius in which another group of Heruls moves there to ask the "royal house" for a new leader. Interestingly, other groups were also present in Illyricum and others even reached north-west Spain via sea in 455[356].

Meanwhile, the Ostrogoths, settled in the Balkans, under the leadership of Theoderic the Amal, were granted lands south of the Danube by the Emperor Zeno of the Easter Roman Empire, but they rebelled in 486, raiding Thrace and threatening Constantinople[357]. The Eastern Emperor, therefore, then came to an agreement with Theoderic: the Ostrogoths thus moved from the Balkans to the Italian peninsula, which meant more independence from the Eastern Empire and a more secure settlement for his warriors and their families[358] . According to the studies, indeed, the number of the Ostrogoths may have been around 200-250.000 people, including 20-30.000 warriors and their families[359]. Moreover, the occupation of Italy was organized so that every family ruled over a land and was in contact with the Amal court: nonetheless, we don't know the relations between the occupants and the native population[360].

At the time, Europe was thus divided between the Visigoths kingdom in Southern France and Spain, the Ostrogoth one in Italy and the Balkans, the Franks in Northern Gaul and Western Germany and the Burgunds, once *foederati* and then masters of their own kingdom, located in the Upper Rhine.

---

[351] *Ibidem.*

[352] *Dall'Impero Romano a Carlo Magno*, La Storia, La Biblioteca di Repubblica, 2004.

[353] Goffart, W.A., *Barbarians and Romans, A.D. 418-584: The Techniques of Accommodation*, cit.

[354] Dall'Impero Romano a Carlo Magno, cit.

[355] Macbain, B., *Odovacer the Hun?*, Classical Philology, 1983.

[356] James E. Europe's Barbarians AD 200-600, cit.

[357] Gwynn D.M., *The Goths: Lost Civilizations*, Reaktion Books, 2017.

[358] *Ibidem.*

[359] Aimone, M., *Romani e Ostrogoti fra integrazione e separazione: il contributo dell'archeologia a un dibattito storiografico*, Reti Medievali Rivista, 13, 1, 2012.

[360] *Ibidem.*

Then, in 568, another Germanic people moved into Italy: the Lombards. This ancient tribe had moved from the lower Elbe, where Tacitus places them in his *De origine et situ Germanorum,* dated 98 CE; the very first contacts with the Romans, however, took place around 5 CE[361]. According to Paul the Deacon they originated in Scandinavia[362]; from there, during the social changes that took place in *Barbaricum* during the third century, they moved to the Upper Danube and even tried to enter Pannonia[363] at the time of the Marcomannic wars (171-175), after which several tribes were settled in the Empire. From then, their name disappears from the Roman records and their moves are difficult to identify; James Bury thus affirms that they probably wandered somewhere north of the Danube and were perhaps included in the Empire of Attila[364]. Then, with Odoacer's victory over the Rugians (487), the Lombards occupied their territories near *Noricum*; however, two decades later they were subdued by the aforementioned Heruls, whose movements are just as difficult to detect as the Lombards' ones[365]. Forced into the lands between the Danube and the Theiss, they rebelled in 508; the remains of the defeated Heruls were settled in Moesia by the Eastern Roman Emperor and found refuge in the lands of the Gepids, the Lombards' new antagonists. Indeed, in order to counterpoise this population, the Lombards were granted the status of *foederati* by Justinian and settled in Noricum and Western Pannonia[366]. Then, they allied with the Avars, an Asiatic population ethically linked to the Huns; this tribe was determined to continue to move westwards and managed to defeat the Gepids in 567[367]. Soon afterwards, the Lombards made another agreement with the Avars: if the Lombards would have conquered Italy, their lands would have been inherited by the Avars, otherwise, if they failed, they would have returned to them. According to Bury the Lombards were little in number[368], and this probably both explains why they decided to move and why they brought with them a Saxon contingent with their families (20.000 people)[369]. Interestingly, the Saxons had been their ancient neighbours, settled in Lombards' original homeland – the upper Elbe – and they had probably maintained contacts with them, even if, once they had conquered Italy, not wanting to submit to their allies, they decided to go back north[370]. The Lombards, on the other hand, never fully conquered the peninsula, probably due to their small numbers and even because of their lack of skills in building a fleet; even though they had lived at the mouth of the Elbe, they were not a maritime people[371].

Sometimes, the conquest of Italy (568) is considered the end of the Migration Period[372]; this date could certainly work as a watershed date in the sense that no other

---

[361] Tacito, *La Germania,* I classici del pensiero libero Greci e Latini, BUR, Milano 2012.
[362] Paolo Diacono, Storia dei Longobardi, Edizione Italiana, Foro Barbarico, cit.
[363] Bury J.B., *The Invasion of Europe by the Barbarians*, cit.
[364] *Ibidem.*
[365] *Ibidem.*
[366] *Ibidem.*
[367] *Ibidem.*
[368] *Ibidem.*
[369] *Ibidem.*
[370] *Ibidem.*
[371] *Ibidem.*
[372] Halsall, Guy. Barbarian Migrations and the Roman West, 376–568, cit.

Germanic tribes moved after that year. On the other hand, in the meantime, another major population movement had begun, that of the Slavs, who continued to migrate and settle in Eastern Europe from the fifth to the ninth century[373].

1.2.1.5 Ending an era: the Avars' and the Slavs' migration

In 558 an embassy arrived in Costantinople: as referred by the historian Theophanes, the whole city gathered to see the newcomers, the Avars, dressed like Huns but with very long hair, tied with ribbons[374]. According to the genetic studies, indeed, this new population was coming from the Far East, like the Huns[375], and was "in flight from its country", in Theophanes's words[376]. The Avars were pleading for peace, and they came to an agreement with Emperor Justinian, who managed to ally with them against the other northern tribes[377]. They had been tributaries of the Turks, but they were probably fleeing from them, taking advantage of their engagement in a war against the Hephthalites. In 561/2 they were offered the lands of Pannonia by Justinian, but not to settle in the Empire as *foederati*. Then, in 567, the Avars made the abovementioned agreement with the Lombards and managed to gain control of the Carpathian Basin, which they would retain until the ninth century. Then, after years of expansion in this region, they would be defeated by the Franks, and the remnants of this once huge population would be absorbed in the Hungarian State[378].

The Avars' evolution is profoudly linked with the one of the Slavs; in fact, the first mention of the latter population predates the Avars: Procopius claims that once they settled in 508 in Pannonia, the Slavs contacted their royal family on the island of Thule. Later, in 537, they are mentioned among the Byzantine reinforcements and again, they are mentioned with the Lombards in the Carpathian Basin but are not precisely located[379]. On the other hand, Jordanes, in his Getica, argues that they came from north of the Vistula[380]. Actually, the archaeological findings are confirming that these tribe probably originated in the area between the Pripyat' basin and the Baltic, but when the Goths moved due to the Hunnic invasion, they also left Eastern Europe[381]. In the sixth century we can identify several archaeological cultures with the Slavs, which we define as "Early Slavs"[382]. From the lower Danube basin,

---

[373] Kazanski, M., *Archaeology of the Slavic Migrations,* Encyclopedia of Slavic Languages and Linguistics Online, 2020.

[374] Pohl W., *The Avars: A Steppe Empire in Central Europe, 567-822*, Cornell University Press, Ithaca, New York, 2018.

[375] Neparáczk, E., et al., *Y-chromosome haplogroups from Hun, Avar and conquering Hungarian period nomadic people of the Carpathian Basin*, Scientific Reports, Sci.Rep.9,1, 16569, 2019.

[376] Pohl W., *The Avars: A Steppe Empire in Central Europe, 567–822,* cit.

[377] *Ibidem.*

[378] Neparáczk, E., et al., *Y-chromosome haplogroups from Hun, Avar and conquering Hungarian period nomadic people of the Carpathian Basin*, cit.

[379] Pohl W., *The Avars: A Steppe Empire in Central Europe, 567–822,* cit.

[380] *Ibidem.*

[381] Kazanski, M., *Archaeology of the Slavic Migrations,* cit.

[382] Dolukhanov P. *The Early Slavs: Eastern Europe from the Initial Settlement to the Kievan Rus,* Taylor and Francis, ebook, 2014.

they moved to the Carpathian Basin and then southwards again, in modern-day Transylvania, around 568 CE. A century later another Slavic population moved in this region and another one in the eight century; probably, these migratory waves arrived under the control of the Avars[383] . The Slavs indeed, had started to raid the Eastern Balkans first, but the Avars quickly took control over them once arrived in the area and changed their social organisation: before the arrival of the Avars, they were neither organized in tribes nor they had a strong leadership[384]. Then, following the Avars' arrival, by the end of the century, they had reached Greece, Peloponnese, the Aegean island and Asia Minor. As already mentioned, Soltyasik suggests that the advance and the settlement of the Slavs in the Balkans was favoured by the effects of the Plague of Justinian[385]. Moreover, even though the Slavs are described as fierce people, once arrived in the southern lands they devoted themselves to agriculture[386].

The Slavs had arrived in the Germanic lands during the sixth century – or even before, according to a recent research[387] – and they reached the Elbe estuary in the eight century in the north and Bavaria in the south, where they were stopped by the Franks: the Germanic-Slav border would not change for more than three centuries[388]. The Slavs had taken the lands left by the Germans, but also the Balkans; the Emperor Justinian was not able to redirect them to westwards, as the Byzantines had done with other tribes, and tried to contain them by appointing them *foederati*[389]. The different Slavic tribes would have later formed the backbone of the States that emerged in the following centuries in Eastern Europe: there, they would meet another Germanic Population that would give the name to the following historical period, the Norsemen[390]. Indeed, the conventional date which closes the Migration Period is 793, the year of the first incursions of the Vikings in England[391].

1.2.2 Explaining people's movements: what caused the Migration Period?

Several hypotheses are usually made to explain the migration of people: research usually distinguish between "push" and "pull" factors, arguing that there are elements that both attract populations into a territory and drive them out of their original homeland[392]. The analysis of the Migration Period, in this sense, must not be an exception to this approach. Nonetheless, the issue is probably complicated by the

---

[383] *Ibidem.*

[384] Soltyasik, K., *The plague pandemic and the Slavic expansion in the 6th-8th centuries,* cit.

[385] *Ibidem.*

[386] *Ibidem.*

[387] Machacek, J., *Runes from L´any (Czech Republic) - The oldest inscription among Slavs. A new standard for multidisciplinary analysis of runic bones,* Journal of Archaeological Science, 127, 2021.

[388] Dolukhanov P. *The Early Slavs: Eastern Europe from the Initial Settlement to the Kievan Rus,* cit.

[389] Procopius, History of the Wars, cit.

[390] Forte, A., Oram, D., et al., *Viking Empires,* Cambridge University Press, Cambridge, 2005.

[391] Barret, J., *What caused the Viking Age?,* Antiquity, 82, 2008.

[392] Schoorl, J., et al., *Push and pull factors of international migration A comparative report,* European Commission, 2000.

relationship between these movements of people and the Roman Empire, as can be seen by the common identification of the so called "Barbarian invasions" as the cause of the fall of Rome[393]. However, new studies are challenging the interpretation for which these movements were the sole cause of the fall of Rome. Moreover, as we have seen throughout this analysis, the pluriannual process of settlement of Germanic tribes as *foederati* makes it difficult to determine a real date for the end of the Roman Empire and some studies even argue that the Caracalla's reform of citizenship in 212 can be seen as a major change that became a relevant factor of attractiveness[394], thus being a kind of beginning of the transformation of Europe into a Roman-Germanic continent. This is demonstrated by the fact that, once arrived in the former Roman Empire, these Germanic people did not apply their rules, but continued to use Latin as the language of culture and of the administration; moreover, Roman culture continued to be a literary and artistic model and Roman law continued to be used as the law of the land. For instance, in the Frankish kingdom the "Roman citizenship" continued to mean "the right to make a will, to give testimony, to buy, to sell"[395], as it was in the Roman period[396]. This could lead us to continue the discussion of the transformation of the Roman Empire, but since we are only dealing with the causes of the Migration Period, we must highlight an element that is becoming evident: the power of the attractiveness that led the Germanic tribes to adopt the laws and the culture of the people whom they had once fought against and over whom they had come to rule. Interestingly, when analysing the area of the Carpathian Basin as the strategic area of departure of the Germanic tribes, the historian Pohl argues that in this frontier zone, during the Migration Period, various military aristocracies were formed. According to the author these groups were looking for the opportunities that the Roman Empire could offer (and the tribal life could not), such as richness, prestige, and power[397]. We can probably extend this analysis to all the border areas between the Roman Empire and *Barbaricum*. In fact, Halsall mentions the role of the Roman Empire in creating the basis for the confederations that arose in the third century: growing internal weakness and instability provided a stimulus to non-Roman rulers which had become dependant on the Roman prestige and gifts and "raised the stakes in barbarian politics"[398]. Therefore, we should see the formation of confederations and alliances like that of

[393] Napier, W., *The Memory Remains: Why the Migration Period and the Fall of Rome Continue to be Mischaracterized as a Barbarian Invasion,* Electronic Theses and Dissertations, 2020.

[394] Ţepelea, M., *Migration and Integration of Barbarians into the Roman Empire,* In Vertreibung, Flucht, Migration, Bern, Switzerland, 2016; Mathisen, R., *"Becoming Roman, Becoming Barbarian": Roman Citizenship and the Assimilation of Barbarians into the Late Roman World,* Migration and Membership Regimes in Global and Historical Perspective, Studies in Global Social History, 13, 2, 2013; Mathisen, R., Peregrini, Barbari, and Cives Romani: Concepts of Citizenship and the Legal Identity of Barbarians in the Later Roman Empire, The American Historical Review, 111, No. 4, 1011-1040, 2006.

[395] *Formulae Arvernenses 27: MGH, Formulae 30, cf. 141, 172, 182, 246, 257-8, 311-3, 518, and throughout* in Mathisen, R., *"Becoming Roman, Becoming Barbarian": Roman Citizenship and the Assimilation of Barbarians into the Late Roman World,* cit.

[396] Mathiesen, R. "Becoming Roman, Becoming Barbarian": Roman Citizenship and the Assimilation of Barbarians into the Late Roman World, cit.

[397] Pohl W., *The Avars: A Steppe Empire in Central Europe,* 567-822, cit.

[398] Halsall, G., *Barbarian Migrations, and the Roman West,* 376-568, cit.

the Alamanni not only as the result of a growing population, as mentioned before, but also as the result of the attraction exerted by the Roman Empire. Which is the perfect definition of a "pull factor".

On the other hand, we should also consider the "push factors". First of all, it is widely agreed that the Huns caused a great change in the European geopolitics of the time: in fact, their arrival pushed the Germanic tribes into the Empire[399]. We have already seen how the Goths begged for permission to enter the Roman Empire, fleeing the Huns. Warriors and their families settled within the borders, after crossing the Danube with all the means possible, including trunks[400]. At the same time, we have also considered the Saxons, Jutes, and Angles as "boat people" leaving their lands rather than invaders, and we also linked the arrival of the Vandals, Alans and Suebi to the Huns' arrival. Interestingly, the revolt that took place after the death of Attila under the leadership of Ardaric is described as the revolt of oppressed peoples; Jordanes, indeed, writes:

> "Ardaric, king of the Gepidae [...] became enraged because so many nations were being treated like slaves of the basest condition, and was the first to rise against the sons of Attila. [...] For by his revolt he freed not only his own tribe, but all the others who were equally oppressed; since all readily strive for that which is sought for the general advantage. They took up arms against the destruction that menaced all and joined battle with the Huns in Pannonia, near a river called Nedao [...] The cause of Ardaric, king of the Gepidae, was fortunate for the various nations who were unwillingly subject to the rule of the Huns, for it raised their long downcast spirits to the glad hope of freedom. Many sent ambassadors to the Roman territory, where they were most graciously received by Marcian, who was then emperor, and took the abodes allotted them to dwell in [...]."[401]

Nonetheless, other factors probably drove the German populations into the Empire. With regard to the neighbouring territories, Halsall claims that internal wars could be the explanation for the migration of tribes such as the Saxons, the Vandals, the Burgundians, the Lombards, and the Heruls, all located in the region behind the frontier peoples[402]. The study of Halsall, *Barbarian Migrations, and the Roman West, 376-568* is a point of reference for understanding the changes that took place outside the Empire; in his opinion the ongoing collapse of the Empire had different effects on the different tribes, forcing some to flee into the Empire, some to seek a future in *Britannia* and some to move southwards in *Barbaricum*[403]. The Elbe-valley, as the Carpathian basin, seems therefore to have played a significant role in the Migration Period, being probably the starting point of a migratory route: the Heruls, the Saxons descending in Italy with the Lombards, and the Lombards themselves, may have followed this route[404]. Moreover, this region also saw the expansion of the Thuringian power: the hegemony of this tribe is perhaps linked to the move-

---

[399] Heather, P., *The Huns and the End of the Roman Empire in Western Europe*, English Historical Review, 110, 435, 1995.

[400] Halsall, G., *Barbarian Migrations, and the Roman West*, 376-568, cit.

[401] Jordanes, *The Origin and Deeds of the Goths*, cit.

[402] Halsall, G., *Barbarian Migrations, and the Roman West*, 376-568, *cit.*

[403] *Ibidem.*

[404] *Ibidem.*

ment of the Lombards and of the Heruls but maybe also due to Frankish pressure[405]. In north-west Germany, there are also signs of a deep crisis, which is linked to the migration of the Saxons to Britain and the formation of the Thuringian confederation. Meanwhile, in the north of this region, in the same period, a new elite began to emerge probably at the expenses of populations such as the aforementioned Heruls; it is probably no coincidence, according to Halsall, that at the time we find the first Danish King, Chlochilaich[406]. In the meantime, great transformations were taking place in Ireland, in society, in politics, and economics; these changes, linked to the movement of the Scots to Britain, probably involved the transition to a new order with respect to the previous provincial kingships[407]. According to Halsall, we can find an explanation in the island's chronicles, in which the outmigrants are considered the exiles from the struggles that took place on the island and which led to the end of several local dynasties. Here the dramatic changes here probably started from the beginning of the fourth century.

To summarise, the new political units were probably on of the "push factor" that drew populations into the political vacuum represented by the collapsing Roman Empire, the "pull factor". Nonetheless, if we broaden the picture, we can see that important changes were taking place in the Southern Mediterranean as well: during the Migration Period, new political entities such as the Islamic caliphates began to appear. The Huns certainly had a role in changing the political situation in *Barbaricum*, but they could have not have an effect on distant populations such as the Arabs – at least, not directly – not to mention the movement of the Scots. Moreover, this process whereby new elites were being shaped by the changed political situation within the Empire was simultaneous in several parts of Europe and began towards the end of the third century[408]: this gives the idea of a global process that could have had several causes. Internal causes including economic and social reason have already been analysed by several research[409] and we won't focus on them; we also dealt with the plague pandemic, and its effect on the society already affected by the political changes that took place during the Migration Period. It is now time therefore, to introduce another variable that could help link these historical changes: climatic variations.

1.3 The Environmental link

1.3.1 The Roman Climate Optimum and its end

The first mention of a possible migration due to environmental causes in Roman times is found in in Strabo's *Geography:* we are told that "Poseidonios" – a 2nd cen-

[405] *Ibidem.*

[406] *Ibidem.*

[407] *Ibidem.*

[408] Alföldy, *The Crisis of the Third Century as Seen by Contemporaries,* Greek, Roman, and Byzantine Studies, 15, 89-111, 1974.

[409] Jones, A.H.M., *The decline and fall of the Roman Empire,* History, 40, 140, 209-226, 2005; Piganiol, A. *The causes of the fall of the Roman Empire,* The Journal of General Education, 5,1, 62-69, 1950.

tury BC philosopher and historian – conjectures that the migration of the Cimbri from their homeland occurred because of an advancing of the sea which did not take place suddenly"[410]. This population lived between modern-day Northern Germany and the Jutland peninsula[411] and became known to the Romans because of their incursion in Roman Italy in 101 BC[412]. Even though Strabo is doubtful about the truthfulness of such event, new insights are revealing that there could be a kernel of truth in his account; not only scientific evidence of floods in the original homelands of the Cimbri has been found, but also archaeological evidence, as the first dams to protect crops were built at the time of the Cimbric migration[413]. Moreover, another research, published by *Nature,* revealed a shift in the North Atlantic Oscillation (NAO) – a meteorological phenomenon characterised by fluctuations in the difference of atmospheric pressure at sea level – which is consistent with a period of drought and famine[414]. A positive NAO oscillation brings more rain to the Northern European region and drought to the Mediterranean, while a negative NAO has the opposite effect, bringing moisture to the southern European region and arid weather in Northern Europe. Interestingly, NAO+ conditions have also been related to increased sea waves heights[415]. One of the earliest Germanic migration could therefore be linked to a change of climatic conditions; in fact the abovementioned research published by Nature went even further, arguing for a possible link between other Germanic migrations and past climate changes[416]. Nonetheless, before going on, we must consider the fact that we lack historical sources for this part of Europe, as already stated for the Justinian Pandemic; we must therefore rely on indirect proof such as dendrological and palynological data or analysis of glaciers. It should be said, however, that in recent year there has been an increasing interest towards the analysis of a possible link between the Germanic migrations and climate change[417].

Even though we are mainly focusing on the Migration Period, it was interesting to mention the Cimbric migration not only because this is one of the first recorded Germanic migration and it can be linked to climate shift, but also because, according to the Historian McCormick's study, it was the last migration before the so-called *Roman climate optimum* or *Roman warm period*, a period of exceptional climatic stability[418]. During this period, which has been fixed between 100 BCE and

[410] Compatangelo-Soussignan R., *Poseidonios and the original cause of the migration of the Cimbri: tsunami, storm surge or tides?* Revue des études anciennes, Revue des études anciennes, Université Bordeaux Montaigne, 2016.

[411] *Ibidem.*

[412] *"Cimbri"*, Encyclopedia Britannica online.

[413] Compatangelo-Soussigna, R., *Poseidonios and the original cause of the migration of the Cimbri: tsunami, storm surge or tides?*, cit.

[414] Drake, B.L., *Changes in North Atlantic Oscillation drove Population Migrations and the Collapse of the Western Roman Empire*, cit.,

[415] Hurrell, J.W., et al., *The North Atlantic Oscillation: Climatic Significance and Environmental Impact*, American Geophysical Union as part of the Geophysical Monograph Series, 134, 2013.

[416] Drake, B.L., *Changes in North Atlantic Oscillation drove Population Migrations and the Collapse of the Western Roman Empire*, cit.

[417] Mc Cormick, M., *Climate Change during and after the Roman Empire: Reconstructing the Past from Scientific and Historical Evidence*, Journal of Interdisciplinary History, 43, 2, 169–220, 2012.

[418] *Ibidem.*

200 CE[419] some regions enjoyed favourable conditions: glaciers retreated and Rome's breadbasket, Egypt, produced cereals in abundance[420]. Nonetheless, it should be said that there is no broad agreement on the fact that this period of climate warming was a global-scale phenomenon[421]. However, studies focusing on Europe essentially agree on the effective prosperity of the European climate during this period; this is confirmed by analyses of ocean sedimentation[422], temperature reconstruction from ice cores, lake productivity and dendrochronological analyses[423], published by authoritative sources such as Nature and Science.

The Roman Warm Period has thus been linked to the emergence of the Roman Empire: between 100 BCE and 200 CE, Rome managed to impose its rule all over the Mediterranean, reaching its maximum expansion in 117, under Trajan[424]. There were no volcanic eruptions during the period between 44 BCE and 169 CE, solar irradiance was elevated, Spain and Italy were well-watered, and the Roman Empire was able to feed every citizen, despite its huge population growth: indeed, no famines were recorded in the high Roman Empire[425]. Thus, given the exceptional climatic stability and the incredible Roman ability to rule all over that immense area, it is tempting to link these two factors. And this becomes even more tempting if we consider that "fall" of the Roman Empire took place during a period of "exceptional climate variability"[426].

Historian Kyle Harper defined the interval between the Roman Warm Period and the Little Antique Ice Age, a period of general cooling, as we will see later – "Roman Transitional period", which he sets between 150 CE and 450 CE[427]. Another research confirmed, indeed, that already in the third century, the frequency of the Nile floods and lake productivity declined, coinciding with the period of political instability known as known Third century crisis[428]. The low frequency of the Nile had important effects on the crops of the Roman breadbasket: evidence of extreme dryness and desertification, abandonment of villages, as well as economic difficulties, irrigation issues and upheavals have been linked to changing climatic conditions during the third and fourth centuries[429]. Moreover, historian Sabine Huebner related the reduced rate of floods of the Nile to a shift of the African Monsoon in the

---

[419] *Ibidem.*

[420] *Ibidem.*

[421] Neukom, R., No evidence for globally coherent warm and cold periods over the pre-industrial Common Era, Nature, 571, 550–554, 2019.

[422] Bianchi, G., McCave, I., *Holocene periodicity in North Atlantic climate and deep-ocean flow south of Iceland,* Nature 397, 515-517, 1999.

[423] Drake, B.L., *Changes in North Atlantic Oscillation drove Population Migrations and the Collapse of the Western Roman Empire,* cit.

[424] Hammond, Mason. "Trajan". Encyclopedia Britannica online, 15 Dec. 2020.

[425] Harper, K., *The Environmental fall of Rome,* Daedalus 145, 2,101-111, 2016.

[426] Bungten, U., et al., 2500 Years of European Climate Variability and Human Susceptibility, Science, 311, 2011.

[427] Harper, K., *The Fate of Rome: Climate, Disease, and the End of an Empire,* cit.

[428] Drake, B.L., *Changes in North Atlantic Oscillation drove Population Migrations and the Collapse of the Western Roman Empire,* cit.

[429] Climate Change in the Breadbasket of the Roman Empire. Reconstructing Nile Floods for the Roman Period, Conference Report, Conference organized by the *Istituto Svizzero di Roma,* 23-24 January 2020

Nile's headwaters region[430]. In fact, after the peak of the second century, records of Nile floods show a decrease (Figure 1) from the third century onwards[431].

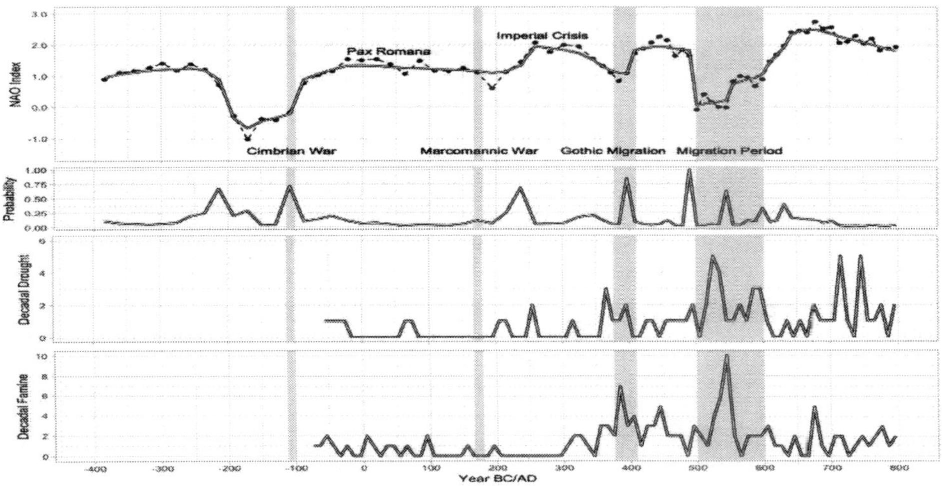

Figure 1: NAO index compared to the droughts and famines in Drake, B.L., *Changes in North Atlantic Oscillation drove Population Migrations and the Collapse of the Western Roman Empire*, cit.

On the other hand, while the Mediterranean region was experiencing these drought phenomena, Northern and Central Europe were under increasing heavy rains, as revealed by the dendrochronological analysis[432]. This was possible due to the different impact of the abovementioned North Atlantic Oscillation Phenomenon, which, as said, has different effects on these two European regions[433]. Meanwhile, the peoples of inner Asia, in the years between 350 and 370 experienced a "multidecadal drought event", as Kyle Harper states[434]. However, the view for which the Huns, coming from inner Asia, would have been driven by that period of megadroughts, also supported by paleo-environment Professor Edward R. Cook[435], has been challenged by Archaeologist Susanne Hakenbeck, who argued that these droughts did not affect the Huns, which were already in Europe at that time[436]; however, they may have influenced the migration of other Asian populations: indeed, according to Priscus, an Eastern Roman diplomat, the Avars left leave their

[430] *Ibidem.*

[431] Drake, B.L., *Changes in North Atlantic Oscillation drove Population Migrations and the Collapse of the Western Roman Empire*, cit.

[432] *Ibidem.*

[433] *Ibidem.*

[434] *Ibidem.*

[435] Cook, E. R., *Megadroughts, ENSO, and the Invasion of Late-Roman Europe by the Huns and Avars*, in The Ancient Mediterranean Environment between Science and History. Leiden, The Netherlands, Brill, 2013

[436] Hakenbeck, S., *The Role of Climatic and Environmental Factors in the Hunnic Phenomenon in South-east Europe*, SocArXiv, 2019.

country due to mysterious "great clouds and mists"[437]. Interestingly, Kyle Harper argues that the said great drought had similar effects to the American "Dust Bowl"[438]. McCormick also agrees that the drought of the fifth century may have led to Asian populations such as the Hun Hephthalites Huns to invade the Sassanian Empire and links the sixth century dry conditions in the Eurasian steppe to the Avars' movement[439].

While disagreeing on the possible effects of climate change in Asia on the Huns, Hakenbeck argues that a series of very dry summers and an increase in aridity occurred in the early decades of the fifth century in the Carpathian Basin; in her opinion, this might have led to a socio-economic reorganization of the Huns, pushing them to demand gold from the Roman provinces, similarly to what had happened with the Xiong-Nu and the Chinese Empire[440]. This hypothesis seems to be confirmed by the analysis carried out by University of New Mexico researcher Lee Drake and published by Nature, which confirmed a positive North Atlantic Oscillation in 375 CE, also proven by textual evidence referring to droughts and famine; Drake thus link the effects of the climate shift in this region to the migration of the Huns and of the Goths, given the effect of the NAO on the Carpathian basin region[441]. Another study based on palaeobotanical analyses confirmed that the climate of this geographical area, climate had been warm during the Roman Climate Optimum, warm enough to cultivate peaches and figs (and probably even vine); however, things changed with the Migration Period[442]. Indeed, from the late fourth century – thus consistent with the Gothic/Huns migration – temperatures in Carpathian basin fell and climate became dryer[443]. These studies are therefore demonstrating that there were major climate changes in this area, which would later be crucial for the future of the Empire. We must therefore focus on the other regions bordering the Empire too.

We have already mentioned the increasing level of precipitation in modern-day France and Germany (see Figure 2); moreover, isotopegeochemical methods applied on peat deposits such that of Durres Maar, in north-western Germany, confirmed that the period from the fourth and seventh century was a cold and wet[444]. Again, a cooler climate is confirmed by dendrochronological and fossil analysis carried out in Northern Finland and Sweden: here temperature decreased from the half of the third century, with minimums in the first decades of the third century, in

---

[437] Howorth, H.H., *The Avars, The Journal of the Royal Asiatic Society of Great Britain and Ireland, Oct., New Series, 21, 4, 721-810, 1899.*

[438] Harper, K., *The Fate of Rome: Climate, Disease, and the End of an Empire,* cit.

[439] McCormick, *Climate Change during and after the Roman Empire: Reconstructing the Past from Scientific and Historical Evidence,* cit.

[440] Hakenbeck, S., *The Role of Climatic and Environmental Factors in the Hunnic Phenomenon in South-east Europe*, cit.

[441] Drake, B.L., *Changes in North Atlantic Oscillation drove Population Migrations and the Collapse of the Western Roman Empire,* cit.

[442] Vadas, A., Rácz, L., *Climatic Changes in the Carpathian Basin during the Middle Ages. The State of Research,* Global Environment 12, 198-227, 2013.

[443] *Ibidem.*

[444] Moschen, R., et al., Temperature variability at Dürres Maar, Germany during the migration period and at high medieval times, inferred from stable carbon isotopes of Sphagnum cellulose, Clim. Past Discuss., 7, 535-573, 2011.

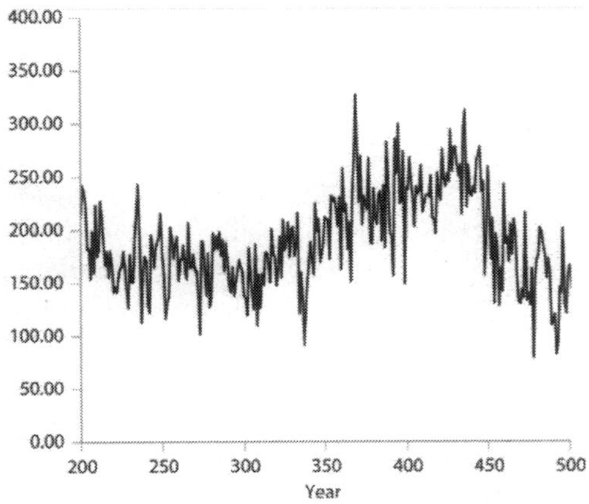

Figure 2: *Precipitation's* data in France and Germany (data from Bungten et al. 2011) in Harper, K., *The Fate of Rome, Climate, Disease, and the End of an Empire*, cit.

the first half of the fourth century, and also during the fifth century and in the sixth century (see Figure 3)[445].

Moreover, according to Lutherbacher et al., European summers in general were colder from 300 to 600 CE, not only in specific regions[446]. This is consistent with the data showing reforestation in many areas of Europe: indeed, palynological studies are confirming that the aforementioned increase in precipitations and lower temperatures were accompanied a decline in agriculture and a general reforestation in north-west and north-central Europe, in northern Poland, in the Baltic States, Southern Sweden, and along the Baltic coast from the fifth century onwards[447]. Increasing human impact in the Eastern Baltic, indeed, is detected again only around the tenth century[448]; furthermore, in North-west and North-central Europe the reversal of the forests, caused by the Roman settlements, stopped after the collapse of the Roman Empire, though this trend is not homogeneous in Europe[449].

Data from South-western Sweden confirm a contraction of agriculture that began almost 1500 years ago; these data are similar to those coming from contemporary *Britannia*[450]. In fact, in his book *The end of Roman Britain*, historian Michael Jones argues that a shift to a colder climate occurred around 400 CE in Britain and also became colder around mid-century: increased rainfall, probably caused floods and aggravated soil erosion, already caused by deforestation and expansion of agriculture[451]. Jones's conclusions are interesting since they probably describe what happened in other parts of Europe too; deteriorating climate conditions and increased precipitations probably led to reduced fertility of fields

[445] Esper, J., et al., *Northern European summer temperature variations over the Common Era from integrated tree-ring density records*, Journal of Quaternary Science, 29, 5, 487-494, 2014.

[446] Lutherbach et al., *European summer temperatures since Roman times*, Environmental Research Letters, 11, 2016.

[447] Gaillard, M., et al., Causes of Regional Change-Land Cover, Second assestment of climate change for the Baltic Basin, The BACC II Author team, Springer Open, Regional Climate Studies, 2015.

[448] Brown, A., Pluskowski, A., *Medieval landscape transformation in the southeast and Eastern Baltic: paleoenvironmental perspectives on the colonisation of frontier landscapes*, Archaeologia Baltica 20, Frontier Societies and environmental change in northeast Europe, Klaipeda, 2013.

[449] Woodbridge-Fisher, J., et. Al., *Vegetation and land-use change in northern Europe during late antiquity: a Regional-scale pollen-based reconstruction*, Late Antique Archaeology, 11, 1, 2015.

[450] De Jong, R., Lageras, P., *Exploring the patterns and causes of land use changes in south-west Sweden*, Veget Hist Archaeobot, 20:15–27, 2011.

[451] Jones, M., *The End of Roman Britain*, cit.

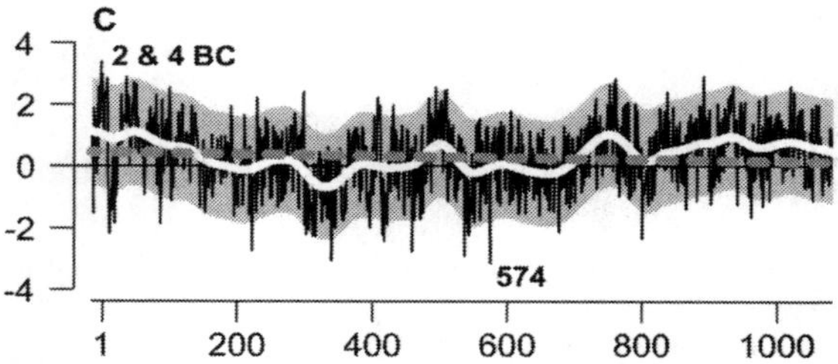

Figure 3: Reconstruction of Northern European temperatures variation from 1 CE to 1000 CE: 2, 4, and 547 are respectively the two maximums and the minimum of the period. In Esper, J., et al., *Northern European summer temperature variations over the Common Era from integrated tree-ring density records, Journal of Quaternary Science* (2014) 29(5), 487-494.

and more bogs; moreover, reduced temperatures may have lowered the elevation level at which grains grows, and may have shortened the growing season[452]. Interestingly, Jones argues that the climate may have been colder in northern regions, thus pushing south populations such as the Picts and the Scots. contemporary and huge migrations into the former Roman Empire to the changing climate conditions; has we have seen throughout this analysis, indeed, raids and circumscribed migrations occurred before the fourth century too, but it is only in the period between the end of the fourth century and the eight century – the Migration Period – that migrations become a large-scale phenomenon, in conjunction with the abovementioned deteriorating climatic conditions. A link between the two phenomena has therefore been suggested by the various abovementioned research[453]. It should be said, however, that due to the said inverse effects of the NAO, the climate was also changing in the Mediterranean region: for instance, decreasing solar insolation in the northern hemisphere and cooler air temperature in Greenland had also effects in Spain. This is proved by a decline in the fluvial influence in the Algerian-Iberian basin, lower lake levels in Spain, cooling events in the Balearic basin, forest regression and decreased river activity in Southern Europe in general[454]. Spatial heterogeneity was, however, a feature of the weather in Southern Europe, as in the same Iberian Peninsula, the northern region experienced humid and cold conditions; two different lakes for instance, reveal extreme dryness and wet conditions in the same period[455]. Nonetheless, as we are focusing on a possible link between Germanic tribes' movements and climatic Change during the Migration period, we will only concentrate on northern and north-eastern Europe, from the tribes came from.

---

[452] *Ibidem.*

[453] Drake, B.L., *Changes in North Atlantic Oscillation drove Population Migrations and the Collapse of the Western Roman Empire,* cit.; Moschen, R., et al., Temperature variability at Dürres Maar, Germany during the migration period and at high medieval times, inferred from stable carbon isotopes of Sphagnum cellulose, cit., Vadas, A., Rácz, L., *Climatic Changes in the Carpathian Basin during the Middle Ages. The State of Research,* cit., Bungten, U., et al., 2500 Years of European Climate Variability and Human Susceptibility, cit.

[454] Nieto-Moreno, V., et al., *Tracking climate variability in the western Mediterranean during the Late Holocene: a multiproxy approach*, Climate of the Past, 7, 1395-1414, 2011.

[455] Sanchez-Lopez, G., *Climate reconstruction for the last two millennia in central Iberia: The role of East Atlantic (EA), North Atlantic Oscillation (NAO) and their interplay over the Iberian Peninsula*, Quaternany Science Reviews 149, 135-150, 2016.

Geography Professor Ellsworth Huntington was one of the first to hypothesise a link between climate change and Germanic migrations in 1907; however, this approach has been lacking in further research, which insisted on other explanations[456]. Only recently has research begun to consider climate variability as one of the possible causes of the the migratory movements during Migration Period. A major contribution was given by Professor of Geography David Holt, who focused its attention on the relation between dendrochronological data – used to reconstruct climate variability – and the major events in *Germania,* that is to say in the region comprised between the Rhine, the Vistula and the Danube; importantly, Dacia was also included in the investigation[457]. Another important study was that of Armin Volkmann, who concentrated on the Oder region. Both studies came to the same results and argued that climatic variability may have been one of the causes of the Germanic tribes' movements. Interestingly, according to David Holt, in the years between 274 CE and 374 CE 59 out of 61 migrations, attacks or settlements in the Roman Empire of the Germanic tribes, occurred in years with low agricultural growth rates: a percentage of 97%. Moreover, the data are consistent with records of droughts in the – few – written documents: for instance, archaeological data from the Netherlands reveal that a cold period ended in 320 CE, and in 348 a multi-year drought occurred in the lands of the Visigoths; both events are confirmed by the dendrochronology analysis (N for Netherlands and A for the Visigoth King, Athanaric, in the figure). On the other hand, it is reported that Dacia experienced a high crop productivity in 357 and two years later the Burgundians became sedentary; again, this is consistent with the prosperous period emerging from the tree analysis (D and B in figure 4).

Volkmann's analysis instead, focused on the Oder region and on the analysis of the settlements. In the early migration period, the climate shifted towards colder and drier conditions, as in other parts of northern Europe[458].

Archaeological evidence indeed shows dramatic processes taking place in this area; while the Roman Period saw intense settlement of the region, there was a decline in the Migration Period, as shown in Figure 5.

Volkmann therefore suggests that the migrations from this area, which began before the cooling period, may have been caused by several factors: first of all, during the climate optimum, by the political weakness of the Roman Empire during the third century. There is evidence, indeed, that individual groups took valuable pieces back to their homeland. Then, the worsening of climate caused the inhabitants to leave their homes, with a peak in migration in the fifth century, as also testified by the sudden high density of the settlements in the Spree-Havel region, an intermediate step in the tribes' southward movement. The contemporary military and political weakness of Rome certainly provided another stimulus – or pull factor – for the Germanic tribes[459] Deprived of the youth – which had already moved southwards –

---

[456] Holt, D.H., *Germania and Climate Variability in 3rd and 4th Centuries A.D.: A Methodological Approach to Dendroclimatology and Human Migration,* Physical Geography, 32, 3, 2011.

[457] *Ibidem.*

[458] Volkmann, A., *Climate change, environment and migration: a GIS-based study of the Roman Iron Age to the Early Middle Ages in the river Oder region,* European Journal of Post-classical Archaeologies, 5, 2015.

[459] *Ibidem.*

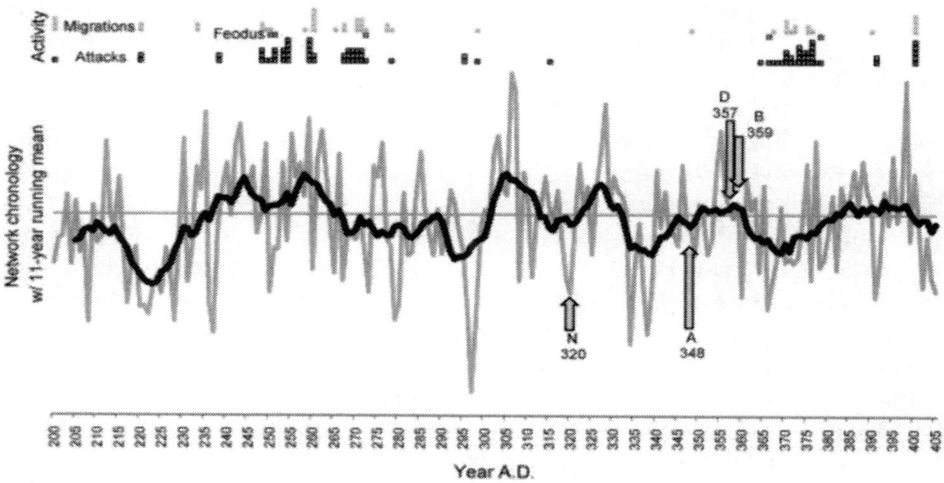

Figure 4: "Activity in and around Germania for the 3rd and 4th centuries A.D. as it compares to the network chronology by year and periodicity. Activity is divided into migrations, feodus (treaties), and attacks, and is ranked by number of each (as per number of squares), with a maximum of six different tribes for migration and five different tribes for attacks. The letters "A" and "N" are recorded droughts and the letters "D" and "B" are recorded prosperous times". In Holt, D.H., *Germania and Climate Variability in 3rd and 4th Centuries A.D.: A Methodological Approach to Dendroclimatology and Human Migration*, Physical Geography, Volume 32, Issue 3, 2011.

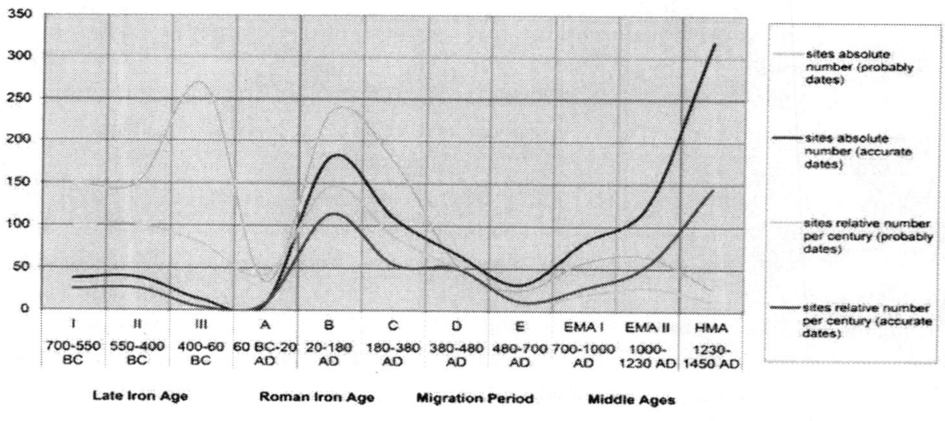

Figure 5: Numbers of sites under investigation in the Oder region. In Volkmann, A., *Climate change, environment and migration: a GIS-based study of the Roman Iron Age to the Early Middle Ages in the river Oder region, European Journal of Post-classical Archaeologies*, volume 5, 2015.

and therefore of the labour force, and faced with. Deprived of the youths and therefore of the labour force and facing the reforestation and the desertification of the agricultural lands, the remaining part of the population decided to leave; interestingly, even though the Oder region may have been a land of contact between the Germanic tribes and the Slavs, there is no evidence that contacts between the two populations occurred, as the land seems to have been deserted for about a century[460]. Volkmann's interpretation has been indirectly confirmed by the study of Lee Drake, which argued that a NAO shift, linked to drought conditions, occurred between 500 and 600 CE, as previously mentioned. Moreover, a settlement contraction is also recorded in the Rhone Valley in the same period: archaeologist Fredric Cheyette showed how the number of settlements in this area significantly decreased from the second century: while there were 565 occupied sites between 100 and 200 CE, this number fell to 224 and 188, respectively in the fourth and in fifth century[461]. Finally, new studies are demonstrating that the Anglo-Saxon migration occurred in a period of increasing sea levels in the North Sea[462]: thus, the loss of lands may have led these people to leave their lands.

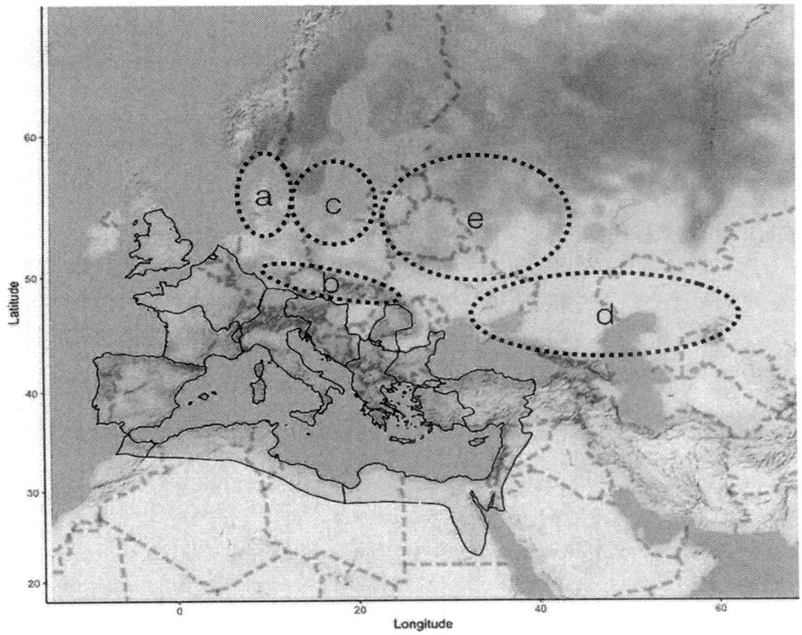

Figure 6: The locations are highlighted for (a) the Cimbri and Teutones before 117 B.C., (b) the Marcomanni and Quadi before 160 A.D., (c) hypothesized location for tribes which would eventually become the Goths before 370 A.D., (d) possible migration path of the Huns around 400 A.D., and (e) Slavic-speaking groups prior to the Migration Period (500 A.D.). In Drake, B.L., Changes in North Atlantic Oscillation drove Population Migrations and the Collapse of the Western Roman Empire, cit.

460 *Ibidem.*

461 Cheyette, F., *The disappearance of the ancient landscape and the climatic anomaly of the early Middle Ages: a question to be pursued*, Early Medieval Europe, 16, 2, 127-165, 2011.

462 Simmons, I.G., *Margins of the East Fen: Historic Landscape Evolution*, Durham University, Appendix 2.1.A Weather, Climate, Sea level and floods in Medieval Times, https://www.dur.ac.uk/east-lincs-history/investigations/the-saxon-shore-the-vikings-and-domesday-book/appendix-2.1.a/

1.3.2 The 536 CE event and the LALIA/DACP

While it is becoming evident that the climate change which occurred in the Migration Period was a general and prolonged climatic shift that took place over several centuries, 536 is remembered as a watershed year. Indeed, contemporary chronicles all across Eurasia make reference to a "dust veil event": Procopius writes that "during this year a most dread portent took place. For the sun gave forth its light without brightness [...] and it seemed exceedingly like the sun in eclipse, for the beams it shed were not clear"[463]. Cassiodorus also claim that the sun was weak:

> "The Sun, first of stars, seems to have lost his wonted light, and appears of a bluish colour. We marvel to see no shadows of our bodies at noon, to feel the mighty vigour of his heat wasted into feebleness, and the phenomena which accompany a transitory eclipse prolonged through a whole year. The Moon too, even when her orb is full, is empty of her natural splendour. Strange has been the course of the year thus far. We have had a winter without storms, a spring without mildness, and a summer without heat. Whence can we look for harvest, since the months which should have been maturing the corn have been chilled by Boreas?[464]"

Now we know that this event was caused by a volcanic explosion, as demonstrated by data coming from glaciers in Greenland and Antarctica[465]; the volcano was probably located in a tropical or subtropical region[466] or in the northern Hemisphere[467] and its effects have been linked to several social changes in the following period[468]. The dust veil was possible because:

> "Dry fogs appear in the atmosphere when large volcanic eruptions inject massive quantities of fine silicate ash and aerosolforming sulphur gases into the troposphere and stratosphere. Although the ash gravitationally settles out within weeks, the aerosols spread around the globe and can remain suspended in the stratosphere for years"[469]

The volcanic eruption of 536 was not the only volcanic event, however, as it was followed by two other volcanic explosions, one in 540 – a tropical volcano, exceeding the force of the Tambora explosion of 1815's – and another one in 547; the general cooling caused by this series of explosion was also sustained by "positive feedback loops of ocean-heat content and sea-ice extension"[470] and by a solar minimum[471].

---

[463] Procopius, History of the Vandalic Wars Volume 2, Books III and IV, William Heinemann London, England, 1916, p. 329.

[464] *The Letters of Cassiodorus,* translation of Thomas Hogdin, printed by Horace Hart, printer to the University, Oxford, 1886 pag. 519.

[465] Larsen, L.B., *New ice core evidence for a volcanic cause of the A.D. 536 dust veil,* Geophysical Research Letters, 35, 2008.

[466] Stothers, R.B., *Mystery cloud of A.D. 536,* Nature, volume 306, 1984.

[467] Peregrine, P., *Climate and social change at the start of the Late Antique Little Ice Age,* The Holocene, 30, 11, 1643-8, 2020.

[468] *Ibidem.*

[469] Stothers, R.B., *Mystery cloud of A.D. 536,* cit.

[470] Bungten, U., et al. *Climate and societal change during the Late Antique Little Ice Age from 536 to 660,* Nature Geoscience, 9, 231-236, 2016.

[471] *Ibidem.*

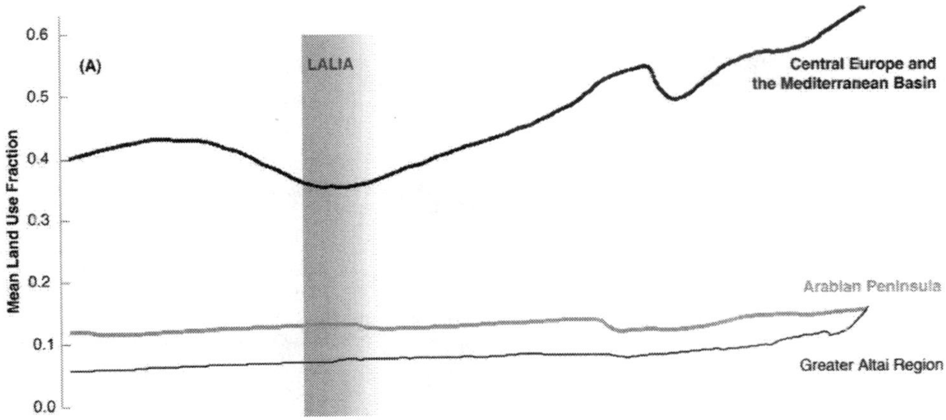

Figure 7: The mean land use in Central Europe and in the Mediterranean Basin. In Bungten, U., et al. Climate and societal change during the Late Antique Little Ice Age from 536 to 660, Nature Geoscience, 9, 231-236, 2016. We can see a general contraction corresponding to the period between 400 CE and 800 CE.

Thus, some authors refer to the period of general cooling as the "Little Antique Ice Age (LALIA)", although there is no broad agreement on this definition or the duration of this cooling phase[472]. Indeed, this period of general cooling has also been identified as "Dark Age Cooling Period (DACP)"; however, both the LALIA and the DACP overlap with the Migration Period but while the LALIA has been set between 536 and 660[473], the term DACP describes the period between 400 and 700[474]. We can therefore consider the LALIA as a period of extreme cold weather within a period of general cooling.

As mentioned above, several changes took place during this period: major upheavals were recorded in central Asia in 550, probably driving the Avars in Europe, which, in turn, could have possibly pushed the Slavs and the Lombards further into Central Europe[475]. The Slavic dialect, indeed, spread all across Easter Europe during the sixth and seventh century and the forces driving westwards this population have not been fully explained[476].

A link between social change and general cooling has also been postulated by anthropologist Peter Peregrine, who created an index, the *Social Change Index* which was compared to the intensity of local cooling; the questions used to determine the

[472] Haldon, J., et al., *Plagues, climate change, and the end of an empire: A response to Kyle Harper's The Fate of Rome (1):* Climate, History Compass, Wiley Online Library, 2018; Helama et al., *Limited Ice Age Cooling,* Nature Geoscience, 10, 242-3, 2017.

[473] Bungten, U., et al. *Climate and societal change during the Late Antique Little Ice Age from 536 to 660*, cit.

[474] Helama S., et al., *Dark Ages Cold Period: A literature review and directions for future research,* The Holocene, Volume 27, 1600-1606, 2017.

[475] Bungten, U., et al. *Climate and societal change during the Late Antique Little Ice Age from 536 to 660*, cit.

[476] *Ibidem.*

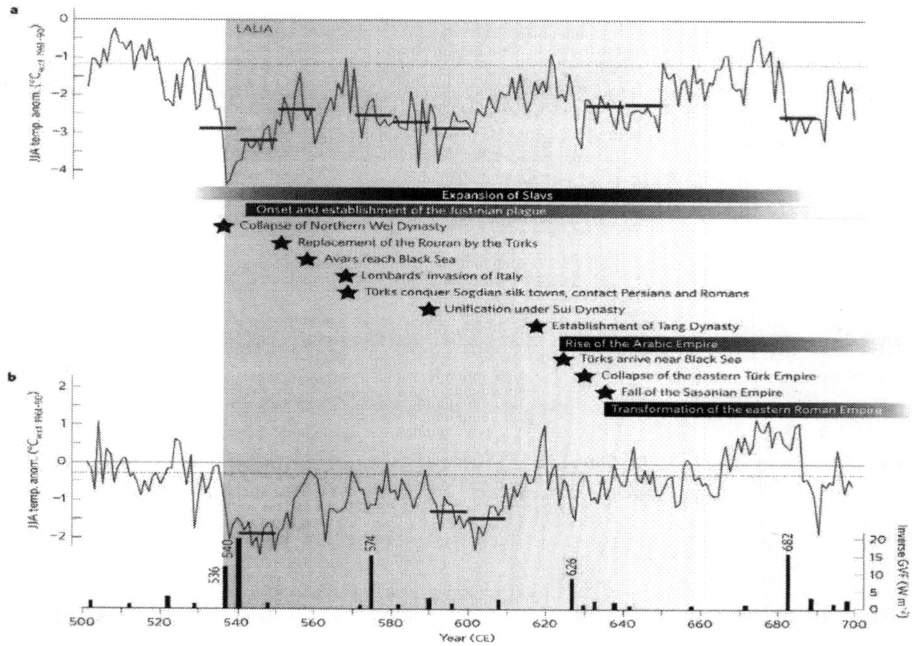

Figure 8: "Reconstructed summer temperatures from the Russian Altai (a) and the European Alps (b), together with estimated volcanic forcing (c)". Blue lines highlight the coldest decades of the LALIA. In Bungten, U., et al. *Climate and societal change during the Late Antique Little Ice Age from 536 to 660*, cit.

degree of the SCI were: "Was there a change in population size or migration? Was there a change in the frequency or intensity of famine or disease? Was there a change in the frequency or intensity of conflict? Was there a change in the organization of communities? Was there a change in the regional organization of the society? Was there a change in the religious or ritual organization of the society?"[477] Then, he compared the results with the temperature change as seen in figure 10.

In his opinion, results proved a link between the temperature change and social change, though they were not "predictability associated with social change among those societies that experienced the most significative cooling"[478]. Therefore, he argues that local variations in effects of the three abovementioned volcanic explosions may have played a role: after further adjustments, he claims that there could be a predictable association between the two variables[479].

Furthermore, as said by Cassiodorus, the 536 event (and the following volcanic explosions) had consequences on the cultivations; this was already evident in the same year, as the event occurred in spring[480]. Cassiodorus indeed states that

[477] Peregrine, P., *Climate and social change at the start of the Late Antique Little Ice Age*, cit.
[478] *Ibidem*.
[479] *Ibidem*.
[480] *Ibidem*.

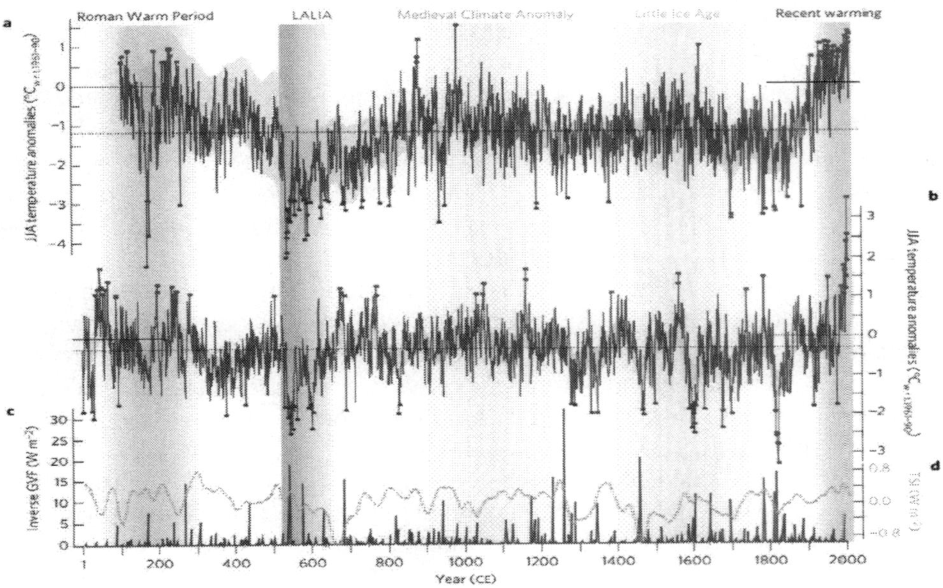

Figure 9: Eurasian summer temperature variability. A and B describe the Altai anc European temperatures, while C and D respectively describes Ice-core-derived global estimates of volcanic and total solar irradianc (in light grey). In Peregrine, P., *Climate and social change at the start of the Late Antique Little Ice Age*, cit.

Case name (short name)	Time frame	Temperature change (degrees C)	Social change index
Europe			
Early Merovingian (Paris)	486 CE–543 CE	−0.79*	6
Ostrogothic Kingdom (Latium)	489 CE–554 CE	−0.69*	14*
Migration Period (Jutland)	500 CE–700 CE	−1.00*	10*
Brega (Ireland)	Reign of Túathal Máelgarb 533 CE–544 CE	−0.51*	6
Toledo (Spain)	Reign of Theudis 531 CE–548 CE	−0.58*	15*

Figure 10: Peregrine, P., Climate, and social change at the start of the Late Antique Little Ice Age, cit.

the this caused the "reaper to fear a new frost in harvest, making the apples to harden when they should grow ripe, souring the old age of the grape-cluster"[481]. This was also remarked by other authors such as the ones of the Irish annals, who recorded "failure of bread"[482]. Indeed, in the *Liber Pontificalis* we read that there was a famine in the "whole world" in 537[483]. Moreover, according to Samuli Helama et al., regardless of the temperature differences, the dust veil affected the

[481] The Letters of Cassiodorus, cit.

[482] *The Annals of Ulster,* Electronic edition compiled by Pádraig Bambury and Stephen Beechinor, CELT, https://celt.ucc.ie//published/T100001A/index.html

[483] Tvauri, A., *The impact of the climate catastrophe of 536/7 AD in Estonia and neihbouring areas,* Estonian Journal of Archaeology, 18, 1, 30-56, 2014.

photosynthesis process, with important effects on the agriculture and, therefore, on societies[484]. Therefore, we will now try to detect the effects on the European peoples of the time.

As Archaeologist Gundersen reports, senior research Stamnes demonstrated that the decrease of just one Celsius degree may have had considerable effects on the cultivations in Northern Europe[485]. Indeed, palynological data referring to the Migration period reveal decreases in arable land in Denmark; in Sweden, a reduction of human activity is detected in the province of Halsingland, in the island of Oland and in the northern regions, while a settlement reduction has been noted around Lake Malaren[486]. In the Uppland region, Archaeologists Price and Graslund observed a large-scale abandonment of settlements: 75% of the villages became deserted and were no longer inhabited, while the percentage of abandonment rises to 92% in the neighbouring regions[487]. Interestingly, these were settlements that had been there even for thousands of years; the two researchers found that some of these settlements were founded again in higher places, but they did not become bigger[488]. Gundersen, however, argues that while in Northern Norway there is no sign of stagnation, this is more evident in Southern and Western Norway[489]. According to Professor of Geography Widgren this is due to a less dependency on agriculture given the abundance of maritime resources[490]. On the other hand, in Southern and Western Norway, the percentage of abandonment is about 90/95% of the previous settlements[491]. Moreover, abandonment of settlements, reforestation, decrease in burials, and changes in material culture are also recorded in Northern Frisia, Schleswig-Holstein, Estonia, Latvia, and Lithuania[492], thus in the whole Northern European region. We previously interpreted this demographic decrease as evidence of the arrival of the plague in Northern Europe, but the number of abandoned settlements could be both a proof of the plague's presence and of outmigration in response to the crop failure. This could be also proved by the fact that in northern Europe, settlement and agricultural reduction were also accompanied by changes in the society: both in Norway[493] and

---

[484] Helama, S., et al., *Volcanic dust veils from sixth century tree-ring isotopes linked to reduced irradiance, primary production and human health*, Scientific reports, Nature, 2018

[485] Gundersen, I.M., *The Fimbulwinter theory and the 6th century crisis in the light of Norwegian archaeology: Towards a human-environmental approach*, Primitive Tider 21, 101-120, 2017

[486] Löwenborg, D. *An Iron Age Shock Doctrine – Did the AD 536-7 event trigger large-scale social changes in the Mälaren valley area?* Cit.

[487] Price N. and B. Gräslund B., *Excavating the Fimbulwinter? Archaeology, geomythology and the climate event(s) of AD 536* In: Past Vulnerability. Volcanic eruptions and human vulnerability in traditional societies past and present. Riede F., 109-132. Aarhus University Press, Aarhus, 2015.

[488] *Ibidem*.

[489] Gundersen, I.M., *The Fimbulwinter theory and the 6th century crisis in the light of Norwegian archaeology*, cit.

[490] Widgren, M., *Climate and Causation in the Swedish Iron Age: Learning from the Present to Understand the Past*, Geografisk Tidsskrift – Danish Journal of Geography, 126-134, 2012.

[491] Price N. and B. Gräslund B., *Excavating the Fimbulwinter? Archaeology, geomythology and the climate event(s) of AD 536*, cit.

[492] Tvauri, A., *The impact of the climate catastrophe of 536/7 AD in Estonia and neihbouring areas*, cit.

[493] Gundersen, I.M., *The Fimbulwinter theory and the 6th century crisis in the light of Norwegian archaeology: Towards a human-environmental approach*, cit.

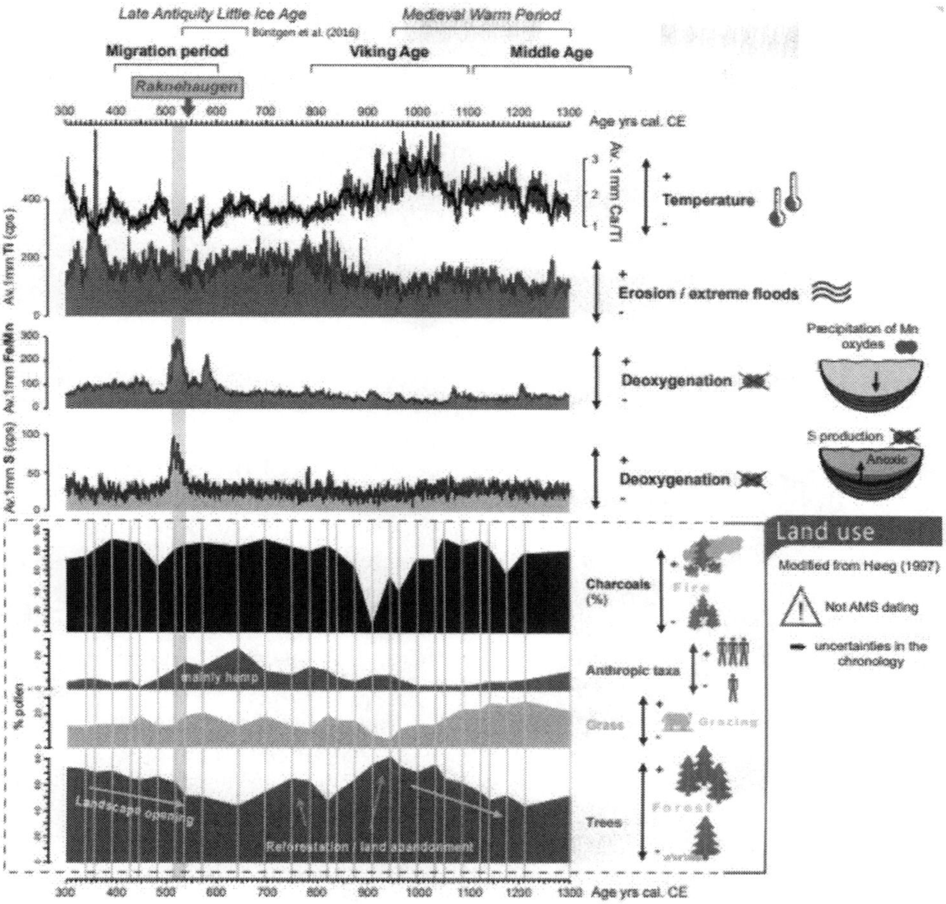

Figure 11. Bajard, M., et al., *Socio-environmental dynamics and volcanic eruptions in the 500-1250 CE period in Scandinavia*, cit. We can see the date of building of Raknehaugen compared to the climatic changes recorded in the lake Ljøgottjern sediment sequence.

in Sweden[494] monumental mounds appeared and there are indications of stronger upper classes[495]. The most important burial mound, Raknehaugen, the largest in all of Scandinavia, was built between 533 and 551 (see figure 10)[496]. Changes in burial habits in Finland are also being studied[497]. In Oland, there is evidence of construction of ringforts and social tensions[498]. Moreover, Price and Berglund also detected a shift

[494] Löwenborg, D. *An Iron Age Shock Doctrine – Did the AD 536-7 event trigger large-scale social changes in the Mälaren valley area?*, cit.

[495] *Ibidem*

[496] Price N. and B. Gräslund B., *Excavating the Fimbulwinter? Archaeology, geomythology and the climate event(s) of AD 536*, cit.

[497] Tvauri, A., *The impact of the climate catastrophe of 536/7 AD in Estonia and neihbouring areas*, cit.

[498] *ibidem.*

in the relationship with the sun: as agrarian societies, the Northern ones were intimately linked with our star, as also witnessed by the several jewels and artistic productions; however, after two millennia, the use of sun symbols comes abruptly to an end in the sixth century[499]. This also happened in Estonia and Finland, where objects decorated with solar symbols had arrived in the third century[500]. In general, this region witnessed a complete and abrupt change in cultural habits; another shift will take place in the thirteenth century[501], i.e. another period of climate change[502].

Interestingly, according to an hypothesis first put out time by archaeologist Martin Axboe, the huge number of gold hoards found in Scandinavia from the half of the sixth century must be related to the disappearance of the Sun: the findings indeed include high-status objects, valuable, and expression of one's status[503]. Therefore, according to Axboe, they must have been deposited for a collective reason[504]; this could also be related to the construction of new burials, as a way in which the elites may have wanted to reassert their power during a crisis. Thus, in Axboe's opinion, the event that triggered such deposits could have been the 536 CE dust veil, given the effects of a darkening of the sun in Northern Europe. He therefore hints to the possibility that this event might have remained in the collective memory under the myth of Ragnarök[505]. This theory was later reiterated by Price and Graslund; in their view, some of the poems that compose Snorri's *Edda* can be dated up to the fifth century[506]. Interestingly, in this period not only the sun symbols disappear, but other symbols, related to the later Norse mythology, replace them; this makes the two authors argue that changes may have occurred in the relationship with the sun – and therefore in the cosmologic conceptions of the people[507]. According to the two authors, this seems to be reflected in the late mythology, where we find reference to a mighty winter, the *Fimbulwinter,* when:

> "snow drives from all quarters, the frosts are so severe, the wind so keen and piercing, that there is no joy in the sun. There are three such winter in succession, without any intervening summer. But before these there are three other winters, during which great wars rage all over the world.[508]"

Moreover, in the Seeress's Prophecy, another Nordic poem, we are told that during Ragnarök, the sun will get dark, the stars will disappear, and the land will be covered by the sea[509]. In addition, it is also said that a huge wolf, called Moongarm, will swallow the moon and cover the sky with blood; then "the sky will get darker, the winds

---

[499] *Ibidem.*

[500] Tvauri, A., *The impact of the climate catastrophe of 536/7 AD in Estonia and neihbouring areas,* cit.

[501] *Ibidem.*

[502] Brooke, J., *The Global Dark and Middle Ages, AD 542-1350,* in Climate Change and the course of Global History – A rough journey, Cambridge University Press, Cambridge 2014.

[503] Axboe, M., *The year 536 and the Scandinavian gold hoars,* Medieval Archaeology, 43, 1999.

[504] *Ibidem.*

[505] *Ibidem.*

[506] Gräslund, B., Price, N. *Twilight of the gods? The 'dust veil event' of AD 536 in critical perspective,* cit.

[507] *Ibidem.*

[508] Sturluson S. *The Younger Edda; Also called Snorre's Edda, or The Prose Edda*, Perlego, 2006.

[509] *Il Canzoniere eddico,* a cura di Piergiuseppe Scardigli, Garzanti Editore, Milano, 2004.

will grow wild [...] sunshine grows black, the summer thereafter, all weather get fickle"[510]. Interestingly, these descriptions are consistent with the weather conditions that the ancient inhabitants of Scandinavia might have experienced[511]: cold summers, cold winters, famine, but also blood-red sunsets, compatible with a major volcanic eruption, just like the one that caused the "fickle weather" in the years following 536 CE[512]. Moreover, the mention of "great wars" might be a reference to the social unrest of the Migration period and which can be witnessed by the traces of upheavals recorded in Oland, as mentioned above. Besides, according to Professor of Quaternary Geology Berglund, the Migration Period was also a period of "rising lake levels and increasing bog growth"[513], whence probably the reference to sea covering the earth. Interestingly, Price and Graslund also note that in Kalevala, a Finnish poem which gathers material probably dating also to the Migration period, we find a figure, the Mistress of Northern Lands, which captures both the sun and the moon, preventing them from shining for several years; the authors claims that the phrase "strange to the birds of the sky" is consistent with a typical reaction to a sun eclipse[514]. However, as in Kalevala the sun and the moon are freed, in *Edda* Ragnarök is a myth of regeneration rather than a myth about the end of the world; this could be explained by the fact that after such difficulties, the cooling period ended and life returned to normality; the event thus became a myth to be remembered and that is how it has come down to us, stressing the importance of climate for the Nordic society of the time.

Conclusion

In the course of this analysis, we have focused on the first plague pandemic and the migratory movements from *Barbaricum*, their effects on a changing society and how they interacted with each other. In the light of the current Covid-19 pandemic and the increasing waves of migration, this analysis becomes timely in trying to understand how the situation in Europe might evolve. Moreover, considering new research, which, as we shall see, highlights links between climate change, migration and pandemics, some scholars have tried to investigate a possible link between past climate change and the outbreak of Justinian's Pandemic. Ongoing studies have focused, for example, on volcanic eruptions, rodent behaviour and flea behaviour in relation to changing climatic conditions[515]. The results, however, have not reached a definitive conclusion. Similarly, a link between climate change and migration cannot be postulated with certainty, although, as Zhang points[516], pre-modern societies

---

[510] Sturluson S. *The Younger Edda; Also called Snorre's Edda, or The Prose Edda*, cit.

[511] Gräslund, B., Price, N. *Twilight of the gods? The 'dust veil event' of AD 536 in critical perspective*, cit.

[512] Price N. and B. Gräslund B., *Excavating the Fimbulwinter? Archaeology, geomythology and the climate event(s) of AD 536*, cit.

[513] Berglund, B., *Human impact and climate changes-synchronous events and a causal link*, Quaternary International, 105, 1, 7-12, 2003.

[514] *Ibidem*.

[515] Gage, K.L., et al., *Climate and vectorborne diseases*, American Journal of Preventive Medicine, 35, 5, 436-50, 2008.

[516] Zhang, D., *The causality analysis of climate change and large-scale human crisis*, Proceedings of the National Academy of Sciences - PNAS 108, 42, 17296-17301, 2011.

were directly influenced by climate. This was also evident from the example of the Ragnarök myth. However, although this analysis is an attempt to find common trends between the two periods, it is also necessary to look at the Migration period with different eyes from modern ones. Climate change, although epochal, was nonetheless a gradual change, bringing about important but nonetheless gradual social changes. The migrations of the Germanic peoples had already begun in the 3rd-4th centuries, when the climate began to change. Throughout history, climate change has led to the displacement of populations in a natural way. Man has always migrated and will always migrate, and this is a lesson we must learn in the face of the migratory waves of our time. Of course, we also need to understand what makes people migrate and how we can act in the face of migratory phenomena, which undoubtedly cause social changes. Similarly, it is useful to analyse a past event such as Justinian's pandemic in order to understand human behaviour in the face of such an event.

Furthermore, another important lesson to be learned from the Migration period is that we must see the 'end' of the Roman Empire more as a transformation than a sudden change. And that we must consider the Germanic 'invasions' more as migrations motivated by a number of factors, including climate change.

The aim of this analysis, in fact, was to understand how Justinian's pandemic and the Germanic migrations interacted with each other, how they changed society and how the people of the time reacted. We have already seen important parallels with the contemporary world, which we will leave, however, to the third chapter. With modern eyes, however, we can look at these events and try to find parallels, but also differences. How different, and in what ways, for example, is the current climate change? And if different, can the human response be the same? Furthermore, how does the increased interconnectedness of the world's climate change influence the changes? And what are the factors behind the migrations and pandemics of our century?

II. The Contemporary Era

Pandemics, Migration and Climate change are not only peculiar of our age, as we observed in chapter one. Throughout the *Migration Period*, indeed, these phenomena contributed to changing the society of the time and paved the way for modern Europe. The plague of Justinian, with its death toll and its consequences on agriculture and settlements, heavily impacted early-medieval Europe. Moreover, mass migrations profoundly changed the European society, shifting Europe's centre from the Mediterranean to its geographical core, modern-day Germany, from where all the Germanic people came from. And finally, climate change, which probably influenced both the pandemic and the migrations definitely marked the end of an era and the beginning of another one. Thus, an historical approach would focus on the following issue: can it happen again? Can the current climate change be the underlying factor of both the several pandemics we experienced in these last twenty years and the mass migrations we daily read of in our newspapers? Can we compare the effects of the plague of Justinian to the ones of Covid-19, for instance? What are the forces that trigger the contemporary migrations, and can we compare them to the ones that were at the basis of the Migration Period? Ultimately, are these changes made faster by a globalized society and can we still act on them?

2.1 The 21st century pandemics and epidemics

As we write, the Covid-19 Pandemic is transforming the world's societies and the vaccines are being delivered throughout the globe. As for now, more than 303 million cases have been confirmed, and at least 5 million people died[1]. To the present date, Covid-19 is the second deadliest pandemics of the twenty-first century, being HIV/AIDS the first, with more than 32 million people died so far[2]. However, these two pandemics are not the only ones of our century: in the past 20 years we also witnessed other epidemics/pandemics such as the 2004's SARS epidemic[3], the 2009's H1N1 pandemic, the 2012's MERS epidemic, recently reemerged[4], the 2015's Zika epidemic[5] and the ongoing Ebola epidemic[6].

Given the huge amount of – contemporary – studies concerning these epidemics and pandemics, we will not focus on them from a medical point of view as we

[1] Dong, E., Du, H., Gardner L. *An interactive web-based dashboard to track COVID-19 in real time,* Lancet Inf Dis. 20,5, 533-534, 2020.

[2] UNAIDS, *Global HIV & AIDS statistics — 2020 fact sheet,* accessed on 26/04/2021.

[3] WHO, *Situation Updates – SARS,* Emergencies preparedness, response, accessed on 26/04/2021.

[4] European Centre for Disease Prevention and Control, *MERS-CoV worldwide overview,* accessed on 26/04/2021.

[5] Istituto superiore di sanità, *Zika virus,* EpiCentro, accessed on 26/04/2021.

[6] WHO, *Ebola virus disease,* accesed on 26/04/2021.

did with the Justinian plague: differently from it, indeed, there is no need to carry out an historical investigation looking for (indirect) data to hypothesise the effective diffusion of the disease and its effects on the society. We will therefore briefly introduce the several pandemics/epidemics and we will then focus on their effects.

EVD or Ebolavirus Disease epidemic

Under the term "Ebola" we commonly describe the disease caused by three virus members of the genus *Ebolavirus*[7]; Bundibugyo ebolavirus, Zaire ebolavirus, and Sudan ebolavirus. These three strains are responsible for the large outbreaks that took place in Africa: the first two epidemics were recorded in 1976 in Nzara, Sudan, and Yambuku, Democratic Republic of Congo, near the river Ebola, from which the disease took its name[8]. According to the World Health Organization, fruit bats are "natural hosts"[9] of the virus, which can spill-over to humans through "blood, secretions, organs or other bodily fluids of infected animals such as chimpanzees, gorillas, fruit bats, monkeys, forest antelope and porcupines found ill or dead or in the rainforest"[10]. Then, it can be transmitted between humans through bodily fluids of infected people, directly or via contaminated surfaces[11]. This makes Ebola a zoonosis, that is to say an infection that is transmitted to humans by animals.

Infected people develop malaise, fatigue, and myalgia, gastrointestinal manifestations, nausea, vomiting, and diarrhoea, substantial fluid losses, headache, conjunctival injection, abdominal pain, arthralgia, and bleeding abnormalities (in less than half of affected patients)[12].

Ebola's last outbreak was recorded in N'Zerekore, Guinea, in 2021[13] and it follows the last and largest one, which started in 2014 in the same country and moved across the borders into Sierra Leone and Liberia[14]. In the period between 2014 and 2016, 28.610 cases were recorded in the three countries, with a death toll of, 67%, 47%, and 28%, respectively in Guinea, Liberia, and Sierra Leone[15]. So far, since 1976, there have been outbreaks in Republic of Congo, Sudan, Gabon, Côte d'Ivoire, South Africa, Uganda, Guinea, Liberia, Sierra Leone, Nigeria, Mali, and Senegal[16]. Recently, two vaccines have been approved by the European Medicines Agency (May 2020) and the US Food and Drug Administration (January 2021)[17]; the development of a vaccine against Ebola started in 1970[18].

---

[7] Campion, E., *Ebola,* New England Journal of Medicine, 382, 1832-42, 2020.
[8] WHO, *Ebolavirus disease,* Fact sheet n° 103, 2014.
[9] *Ibidem.*
[10] *Ibidem.*
[11] *Ibidem.*
[12] Campion, E., *Ebola,* cit.
[13] WHO, *Ebola outbreak 2021*- N'Zerekore, Guinea, accessed on 26/04/2021.
[14] WHO, *Ebola virus disease*, Fact sheet, 23 February 2021.
[15] *Ibidem.*
[16] *Ibidem.*
[17] *Ibidem.*
[18] Campion, E., *Ebola,* cit.

The AIDS/HIV pandemic

The first established case of AIDS was located in Kinshasa in 1959, although it was probably not the real first case, as it was discovered thirty years later, by processing random blood samples[19]. The effective beginning of the pandemic can be set in 1981, when the American Centre for disease control *Morbidity and mortality weekly report* announced a new epidemic after studying the anomalous emergence of several cases of a rare infective disease, pneumocystis, in Los Angeles[20]. New outbreaks of this new disease were then recorded in the USA, in Haiti, France, and Western Europe[21].

The disease, the Acquired Immune-Deficiency Syndrome (AIDS), is caused by the Human immunodeficiency virus (HIV) and leads to the decline of the immune system, leaving the organism defenceless against external pathogens[22]. According to the research, HIV's closest relatives are viruses that cause immunodeficiency among chimpanzees (*Pan troglodytes troglodytes*) and western lowland gorillas (*Gorilla gorilla gorilla*) in west central Africa[23]; thus the hypothesis that transmission to human could occur due to hunting, butchering, and eating the meat of chimpanzees[24]. Therefore, as Ebola, AIDS is a zoonosis too. In addition, according to some studies, the virus emerged between the end of the nineteenth and the beginning of the twentieth century in West and Central Africa, where it evolved into a new strain in the 60s and reached Haiti, where it genetically recombined; then, probably between 1969 and 1972, it arrived in the US[25].

As the number of infections was growing, in 1994, the UN announced the creation of the Joint UN Programme on HIV and AIDS, UNAIDS, aimed to provide coordination and technically support between the different stakeholders involved in the research on the disease[26].

So far, more than 75 million people have contracted HIV, and more than 32,7 million people have died from AIDS-related illnesses[27]. Sub-Saharan Africa shares the largest health burden of HIV/AIDS: indeed, while in Europe the number of deaths people is less than one on 100.000, in this part of the world it reaches 200 on 100.000[28]. Globally, AIDS-related deaths peaked in 2005/6 but then began to decline, as did the number of new infections, although, on the other hand, today the number of people living with HIV is the highest on record[29].

---

[19] Iliffe J., *The African AIDS Epidemic: A History*, Ohio University Press, Athens, Ohio, USA, 2005.

[20] Visco, G., Comandini U., Franceschi, S., *AIDS*, Universo del Corpo, 1999.

[21] *Ibidem*.

[22] *Ibidem*.

[23] Sharp, P.M., Hahn, B.H., *The evolution of HIV-1 and the origin of AIDS*, Phil. Trans. R. Soc, 2010.

[24] Rogers, K., Dude, A., Siliciano, R., Dorwick, K., *AIDS*, Encyclopedia Britannica online, 2020.

[25] *Ibidem*.

[26] UNAIDS, *Who we are*, accessed on 27/04/2021.

[27] UNAIDS, *Global HIV & AIDS statistics - 2020 fact sheet*, cit.

[28] Roser, M., Ritchie, H., *HIV / AIDS*, Published online at OurWorldInData.org. Retrieved from: https://ourworldindata.org/hiv-aids, 2018.

[29] *Ibidem*.

So far, notwithstanding the existence of several promising candidates, no vaccine has yet been developed for HIV[30].

The Severe Acute Respiratory Syndrome (SARS) epidemic

The SARS epidemic, caused by a coronavirus, first emerged in China, which recorded the highest number of cases (5327), with a mortality rate of 7%[31]. The first documented case emerged in Foshan, in the province of Guangdong, on the 16th of November 2002, and by January 2003 it had evolved in a major outbreak; the following month new cases were recorded in seven other provinces, and then in Hong Kong, from where it spread to Singapore, Canada, USA and Vietnam[32]. In total, by the end of the pandemic in 2003, 32 countries reported documented cases: 8437 people were infected and 813 died[33].

According to Feng et al., the pandemic was possible because of the limitations of the public health system in China, as functional infectious diseases surveillance was not yet available; this was also complicated by the fact that it was not serologically diagnostics and by the change of the definition of SARS during the epidemic[34].

The illness appeared with fever, aches, dry cough and, in some patients, respiratory difficulties made mechanical ventilation necessary[35]. As Ebola and AIDS, SARS's virus, SARS-CoV, is considered to be of zoonotic origin: this assertion is sustained by the findings of SARS-related coronavirus in masked palm civets as well by the detection of SARS-CoV infections in wet markets where these animals were sold[36].

The "swine flu" pandemic

The "swine flu" was caused by a strain of H1N1 virus similar to those that have been responsible for other flu pandemics such as the ones of 1918/9 (the "Spanish flu"), and of 1977 (the "Russian flu"); it first broke out in Mexico in 2009, then spread to the USA, and from there to the other parts of the world[37]. The clinical symptoms of the "swine flu" were similar to the ones of the normal flu[38]; moreover, the great majority of hospitalized patients had underlying conditions. According to the WHO, more than 214 countries reported outbreaks of the "swine flu" and 18449 casualties were recorded globally[39], though new studies believe that the re-

---

30 European Commission - CORDIS, *A major boost for HIV vaccine identification and development*, EU research result, accessed on 28th of April 2021.

31 WHO, *Summary of probable SARS cases with onset of illness from 1 November 2002 to 31 July 2003*, 24 July 2015.

32 Feng, D. et al., *The SARS epidemic in mainland China: bringing together all epidemiological data*, Tropical medicine & international health, 14,1, 4-13, 2009.

33 *Ibidem*.

34 *Ibidem*.

35 *"SARS"*, Encyclopedia Britannica online.

36 Feng, D. et al., *The SARS epidemic in mainland China: bringing together all epidemiological data*, cit.

37 *"Influenza A H1N1"*, Encyclopedia Britannica online.

38 del Rio C, Guarner J., *The 2009 influenza A (H1N1) pandemic: what have we learned in the past 6 months*, Trans Am Clin Climatol Assoc., 121, 128-140, 2010.

39 WHO, *Pandemic (H1N1) 2009 - update 112*, 2010.

al number of deaths might have been higher (between 100.000 and 400.000 in the first year), along with the number of infected people[40]. The largest death toll may have occurred in South-east Asia and in Africa, that is to say in countries were the access to health care and nutritional status, as well as living conditions, are extremely different from the ones of the other parts of the world[41].

As already mentioned, it is commonly believed that the epicentre of the influenza pandemic was Mexico, where on the 24th of April 2009, the authorities adopted social distancing interventions. Subsequently, when international cooperation between NAFTA health authorities provided further information about the disease and new cases were detected around the world, the WHO raised its status from epidemic to pandemic[42].

According to the American Centre for Disease Control, the virus resulted from "reassortment", that is to say a process in which two or more influenza virus "can swap genetic information by infecting a single human or animal host"[43]. This is believed to have occurred among "North American pig herds and among Eurasian pig herds"[44], changing the previous influenza virus, thus giving individuals little or no protection against this new strain[45].

The MERS epidemic

Another coronavirus, after the one responsible of the SARS, was detected in June 2012 in a Saudi man: the man died of acute pneumonia and renal failure[46]. Soon after, in retrospective, a cluster was identified in a Jordan hospital in April 2012, and in September 2012 three cases were identified in the UK[47]. Other cases were later detected in South Korea, where a single person, returning from the Middle East, started a "nosocomial outbreak" which involved 16 hospitals and 186 patients.[48] As the disease was detected predominantly in countries of the Middle East, it was defined *Middle East Respiratory syndrome*[49]. So far, 2566 cases were confirmed, including 882 associated deaths: death toll has thus been estimated around 35%, even though this could be an overestimation due to possible misinterpretations of mild cases[50]. Twenty-seven countries have been affected so far[51].

---

[40] Istituto superiore di sanità, *Influenza pandemica A(H1N1)pdm09*, 2015.

[41] Centres for Disease Control and Prevention, *First Global Estimates of 2009 H1N1 Pandemic Mortality Released by CDC-Led Collaboration*, 2012.

[42] del Rio C, Guarner J. *The 2009 influenza A (H1N1) pandemic: what have we learned in the past 6 months*, cit.

[43] CDC, *Origin of 2009 H1N1 Flu (Swine Flu): Questions and Answers*, 2009.

[44] *Ibidem*.

[45] *Ibidem*.

[46] De Wit, E., et al., *SARS and MERS: recent insights into emerging coronaviruses*, Nature reviews, microbiology, volume 14, 2016.

[47] *Ibidem*.

[48] *Ibidem*.

[49] Rogers, K, "*MERS*", Encyclopedia Britannica, 2017.

[50] WHO, *Middle East respiratory syndrome coronavirus (MERS-CoV)*, accessed on 26/04/2021.

[51] De Wit, E., et al., *SARS and MERS: recent insights into emerging coronaviruses*, cit.

Symptomatology includes fever, cough, shortness of breath, and pneumonia; moreover, the virus appears to cause more problems to elderly people or people with weakened immune system, as well as to people with previously detected chronicle diseases[52].

The origin of MERS-CoV has been linked to the dromedary camels[53]: evidence of circulation of the virus among these animals dates back to 1983, when the first cases were documented in Middle East, East Africa and North Africa (among animals)[54]. Even though the presence of both infected dromedary camels and infected humans seems to clarify the spill-over between the two species without the presence of an infected intermediate host, it is still unknown how the disease can effectively transit from camels to humans[55]. Some studies claim that eating camels' flesh or drinking camels' milk might have been the cause of the splill-over[56]. MERS is another example of zoonosis.

The Zika epidemic

Zika was discovered in Uganda in 1947 from the blood of a sentinel rhesus macaque but remained practically obscure for almost seventy years; it was then identified in humans in 1952, but the major outbreak took place in 2007 in Micronesia[57]. Then, in 2015 Zika virus arrived in Brazil and spread throughout the Americas; as it became the first major infectious disease linked to human birth defects, the WHO declared a Public Health Emergency of International concern[58].

It is primarily transmitted by the bite of an infected mosquito belonging to the *Aedes* genus, but in the majority of the cases, its symptoms are mild and include fever, conjunctivitis, muscle pain, malaise and headache[59]. No vaccine is available so far[60].

The Covid-19 pandemic

The current Covid-19 Pandemic is caused by the *Severe Acute Respiratory Syndrome coronavirus 2* (SARS-CoV-2), which presents similar characteristics to SARS-CoV, responsible for the SARS Pandemic in 2004[61]. It was reported for the

---

[52] WHO, *Middle East respiratory syndrome coronavirus (MERS-CoV)*, cit.

[53] Azhar, E., et al., *Evidence for Camel-to-Human Transmission of MERS Coronavirus*, New England Journal of Medicine, 370, 2499-2505, 2014.

[54] De Wit, E., et al., *SARS and MERS: recent insights into emerging coronaviruses*, cit.

[55] *Ibidem.*

[56] Roos, R., *MERS outbreaks grow; Malaysian case had camel link*, Center for infectious Disease Research and Policy, University of Minnesota, 2014.

[57] Kindhauser, M.K., *Zika: the origin and spread of a mosquito-borne virus*, Bulletin of the World Health Organization, 2016.

[58] Petersen, L.R., et al., *Zika Virus*, New England Journal of Medicine, 374, 1552-1563, 2016.

[59] WHO, *Zika virus in the Western Pacific*, accessed on 27/04/2021.

[60] Istituto superiore di sanità, *Zika virus*, cit.

[61] van Doremalen, N., et al., *Aerosol and Surface Stability of SARS-CoV-2 as Compared with SARS-CoV-1*, The New England journal of medicine, 382,16,1564-1567, 2020.

first time in late 2019 in Wuhan, where the first cases were related to the local seafood market[62], where animals like snakes, birds and bats were sold[63]. Symptoms includes fever, cough, headache, fatigue, and nausea, among others; however, disease severity seems associated with already present pathologies[64]. The features of Covid-19 are thus similar to those of SARS and MERS[65].

On the 31st of December 2019, the Who office in China picked up a media statement by the Wuhan Municipal Health system reporting the discovery of a "viral pneumonia"; ten days later the same IO announced that Chinese authorities had determined that the outbreak was caused by a new coronavirus[66]. On the 23rd of January, a travel ban was imposed on Wuhan[67]; then, universities, schools and gathering places were closed and outside activities were limited[68]. These measures would be followed by other countries in the successive weeks[69].

So far (10th of January 2022), more than two years after the first outbreak, 303.383.898 cases of Covid-19 have been reported , including 5.481.888 deaths, while 9.389.586.418 doses of vaccine have been administered[70].

However, we still ignore the origin and the source of the virus responsible for the pandemic[71]; moreover, even though the first detected cases of Covid-19 were related to Wuhan's seafood market, research has shown that patient zero, the first to develop the symptoms, had not been there and show no epidemiological links with the other cases[72]. In addition, the virus shares only less than 80% sequence identity with SARS-CoV, but it's extremely similar (96,2%) to Bat-CoV RaTG13, previously detected in a bat in the Yunnan Province[73]. Moreover, similarities have also been found with coronaviruses hosted in Malayan Pangolins[74].

The final WHO report on Covid-19's outbreak was published on the 30th of March 2021 and concluded that the virus was already circulating in late 2019, thus calling for further analysis of the samples of possibly detected cases dating before January 2020[75].

---

[62] Zhou, P., *A pneumonia outbreak associated with a new coronavirus of probable bat origin*, Nature, 579, 7798, 270-273, 2020.

[63] Gouvea Dos Santos, W., *Natural history of COVID-19 and current knowledge on treatment therapeutic options*, Biomedicine & Pharmacotherapy, 129, 2020.

[64] *Ibidem*.

[65] WHO, *Timeline: WHO's COVID-19 response*, accessed on 27/04/2021.

[66] *Ibidem*.

[67] Hien, L., et al., *The positive impact of lockdown in Wuhan on containing the COVID-19 outbreak in China*, Journal of Travel Medicine, 27, 3, 2020.

[68] *Ibidem*.

[69] *Ibidem*.

[70] WHO, *WHO Coronavirus (COVID-19) Dashboard*, accessed on 27/04/2021.

[71] *WHO-convened Global Study of Origins of SARS-CoV-2: China Part*, Joint Report, 2021.

[72] Huang, C., et al., *Clinical features of patients infected with 2019 novel coronavirus in Wuhan, China*, Lancet, 395, 10223, 497–506, 2020.

[73] Liu, Y., et al., *COVID-19, the first documented coronavirus pandemic in history*, Biomedical Journal, 43, 328-333, 2020.

[74] *WHO-convened Global Study of Origins of SARS-CoV-2: China Part*, cit.

[75] *Ibidem*.

2.1.2 Pandemics 'effects on 21st century Europe' society

In light of a comparison with the Justinian's plague, we will adopt the same approach used for the mentioned pandemic, thus focusing on the demographic, economic, political and social effects of the 21st century's diseases in Europe.

To date, 1.650.836 people died of Covid-19 in Europe[76]. Among the 21st century pandemics and epidemics, therefore, Covid-19 is the one with the most significant impact on Europe; to date, indeed, AIDS related deaths in the period 2009-2019 were 19.484[77]; SARS related deaths were 0[78]; H1N1 deaths "did not reach levels normally seen during interpandemic influenza epidemics"[79]; Ebola cases in the EU were 2 and the associated risk has been defined "very low"[80]; some MERS cases were detected in six European countries but did not spread[81]; finally, the likelihood of local Zika transmission has been defined "moderated"[82]. Therefore, we will mainly focus on the effects of the Covid-19 pandemic.

In order to better asses the demographic effects, we should consider the excess mortality, that is mortality rate that exceed the normal rate[83]. According to data, between January and February, European Union usually records peaks of mortality rate, but in 2020 this rate was actually below the average for all EU countries; then, after the Covid-19 outbreak, peaks in the death rate were recorded in April 2020 (25%) and November 2020 (40.7%)[84]. Between January and February 2021 another peak was recorded (40% higher than normal)[85]. Since the beginning of the pandemic, Spain (79,9% in April 2020), Bulgaria (94,4% in November 2020), Lithuania (81% in December 2020) Poland (97% in November 2020), Slovenia (90% in November 2020) registered the highest excess rates; conversely, Finland's excess rate only exceeded 8% (reaching 8,1%) in April 2020. Similar results were recorded in Denmark (the excess rate went above 7% only in December 2020 and January 2021, reaching 10%). Germany (reached 30% and 22% only in December 2020 and January 2020 respectively, but never exceeded 9% in the other months), Estonia (the highest peak was 13%) and Greece (only twice above 20% but always under 10% in the other months)[86]. Denmark had the lowest excess rate, -12% in March 2021. On the other hand, it is worth stating that,

[76] ECDC, *Cumulative number of coronavirus (COVID-19) deaths in Europe* in Statista, retrieved on January 10, 2021.

[77] ECDC, *Number of deaths among people diagnosed with HIV and/or AIDS in Europe from 2009-2019*, Statista, retrieved on April 28, 2021.

[78] WHO, *Summary of probable SARS cases with onset of illness from 1 November 2002 to 31 July 2003*, 2015.

[79] Gauci, A. et al., *The 2009 A(H1N1) pandemic in Europe, A review of the experience*, Special Report, ECDC, 2010.

[80] WHO, *Ebola outbreak in West Africa and the risk to Europe*, WHO Regional Office for Europe.

[81] WHO, *Middle East respiratory syndrome coronavirus (MERS-CoV) and the risk to Europe*, WHO Regional Office for Europe.

[82] WHO, *Zika virus outbreak and the risk to Europe*, WHO Regional Office for Europe, 2016.

[83] Eurostat, *Excess mortality – statistics*, accessed on 28/04/2021.

[84] *Ibidem*.

[85] *Ibidem*.

[86] *Ibidem*.

both in the UK and in Italy, higher numbers of deaths were only recorded during the Second World War[87].

Differences in death toll have been explained by making references to "trained immunity in the population, early and fast education, rapid organization and adaptation of the hospitals and the public, preparedness for pandemics and public hygiene"[88].

Turning to the economic effect of the pandemic, we must consider that the disease triggered a contraction of the general employment, greater than the one experienced in 2009 during the financial crisis: indeed, five million fewer people were employed in the second quarter of 2020[89]. This was due to containment measure, which restrained the demand of labour and discouraged job search[90].

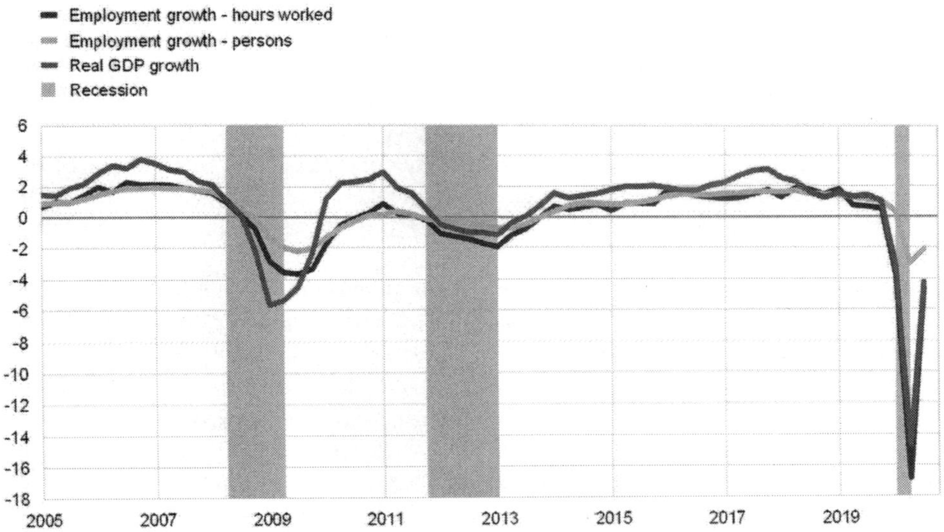

Figure 12: Employment developments in the euro area, Anderton, R. et al., *The impact of the COVID-19 pandemic on the euro area labour market*, ECB Economic Bulletin, Issue 8, 2020

As also evident from the figure, drops in employment rates are highly related to GDP losses[91]; EU's GDP fell by 5,8% in 2020 – and by 6,34% in the Euro area[92] be-

[87] Triggle, N., *Covid: 2020 saw most excess deaths since World War Two*, BBC, 12 of January 2021; Reuters, *Italia, 2020 ha registrato mortalità più alta da Dopoguerra a causa del Covid – Istat*, 5 of March 2021.

[88] Bousquet, J., *Is diet partly responsible for differences in COVID-19 death rates between and within countries?*, Clinical and Translational Allergy, 10, 16, 2020.

[89] Anderton, R. et al., *The impact of the COVID-19 pandemic on the euro area labour market*, ECB Economic Bulletin, 8, 2020.

[90] *Ibidem*.

[91] *Ibidem*.

[92] O'Neill, A., *Growth of the real gross domestic product (GDP) in the European Union and the Euro area from 2016 to 2026*, Statista, 24th of November 2021.

ing the biggest decline following WWII[93]. However, decrease was followed by an unexpectedly faster rebound of 5,1% in 2021[94]. As for international trade, Europe witnessed a contraction in the exports, which decreased less than the imports, thus provoking a trade deficit in April 2020[95]. This was in line with the general international trade contraction[96]. In particular, the average decrease in imports was between -10% and -15%: while Luxembourg experienced the worst import reduction (-43%), Croatia and Slovenia saw an increase in imports (2% and 1%, respectively)[97]. On the other hand, as for exports, the mean range was between -5% and -10% decrease: France, Luxembourg and Greece presented the worst data (-20%, -20% and -19%, respectively), while Slovenia, Estonia and Latvia, the best ones (12%, 6% and 0,05%, respectively)[98].

Retail trade dropped too due to the containment measures: actually, in the very first month of the pandemic, March 2020, food products and sales in supermarket increased, but in the following months food products' sales went down, as the non-food products; in April, total retail trade volume decreased by 11.2%[99]. Decreases in retail trade volume followed the restriction measures: the general level of sales, indeed, went up in May, June, August, October and December 2020, while it went down in April, September and November 2020; in February 2021 the general level was 97.8% of the pre-pandemic level[100]. The total sales volume in the EU is now 3.1 % higher than in February 2020 (but it's "only" 2.6 % higher in the euro area).

Moreover, the pandemic also had effects on the inequality rate: indeed, according to an article published by the European think tank *Bruegel,* Covid-19 had a remarkably different effect on workers with lower educations compared to the ones with tertiary education[101]. However, while jobs for low-educated workers fell everywhere, those for tertiary-educated workers increased in fourteen countries but decreased in eleven; overall, 8% of low-educated workers lost their jobs between the end of 2019 and the first four months of 2020[102]. Meanwhile, jobs requiring university degrees increased by 3%, the same percentage of professional jobs; middle level qualification jobs, on the other hand, decreased by 5%, and so did the number of managers (4%)[103]. Teleworking made the difference too: it is reported that 70% of those who have a university degree are also able to work from

[93] The World Bank, *COVID-19 to Plunge Global Economy into Worst Recession since World War II*, 8th of June 2020.

[94] O'Neill, A., *Growth of the real gross domestic product (GDP) in the European Union and the Euro area from 2016 to 2026*, Statista, 24th of November 2021.

[95] Eurostat, *Impact of COVID-19 on international trade by Member State*, 2021.

[96] The World Bank, *How COVID-19 is changing the world: a statistical perspective Volume II*, Committee for the Coordination of Statistical Activities, World Bank Development Data, 2021.

[97] Eurostat, *Impact of COVID-19 on international trade by Member State*, cit.

[98] *Ibidem.*

[99] European Commission, *Impact of Covid-19 crisis on retail trade,* 2021.

[100] *Ibidem.*

[101] Darvas, Z., *COVID-19 has widened the income gap in Europe*, Bruegel Blog, 3 December 2020.

[102] *Ibidem.*

[103] *Ibidem.*

home, while the percentage drops to 15% for those who did not complete secondary school[104]. Therefore, lower educate people tend to have lower incomes; thus, inequality is rising in Europe, despite the massive social spending[105]. This conclusion is consistent with another study, which argues that major epidemics in this century resulted in income inequality[106]. As for the other data, however, there are major differences among the European countries: the highest working ability during a lockdown, for instance, is higher in countries such as Netherlands, Luxembourg, Sweden and Finland, while it is lower in Bulgaria, Hungary, Slovakia and Cyprus[107]: this clearly affects income inequality.

Moreover, Covid-19 also increased poverty and social exclusion: in particular, the most affected ones were "self-employed workers, vulnerable children, low-income families, Roma, migrants and asylum seekers"[108], but also "elderly people"[109]. As for the latter ones, indeed, the measures of social distancing that were required to ensure the right protection of this category, often resulted in over-isolation: on the other hand, in some cases local epidemics developed in care homes too[110]; to date, according to a survey carried out in Italy, more than 60% of the care homes does not allow any visit or leaving, while no video-calling are available in little less than 50% of the cases[111]. In Finland, too, it is reported that lone elderly people with multiple diseases and physical disabilities were the most vulnerable, thus demonstrating a pan-European trend; indeed, people with disabilities and low income often remained isolated at home, with extremely negative effects[112].

Among the people who are suffering the most for the effects of the pandemic, there are the homeless: during the hardest days of the pandemic, indeed, they lacked the usual assistance and even when they had a place to stay, it was often overcrowded, thus increasing the risk of contagion in already weak people[113]. No access to water or hygienic services was also recorded[114]. Moreover, people with temporary jobs, those who worked outside trade union coverage or self-employed people, being already at risk of losing their jobs before the pandemic outbreak, were among the most hit too: as they were not eligible for the crisis benefits, they experienced financial problems, as well as people with low-income jobs[115]. Before Covid-19, 21% of the Europeans (92,4 million people) were at risk of poverty or social exclusion: Bulgaria, Romania, Greece and Italy scored the highest rates (32,5%,

[104] *Ibidem.*

[105] *Ibidem.*

[106] Furceri, D. et al., *COVID-19 will raise inequality if past pandemics are a guide*, VoxEu, 08 May 2020.

[107] Palomino, J., et al., *Wage Inequality and poverty effects of the lockdown in Europe,* European Economic Review, 129, 2020.

[108] Malgesini, G., *The impact of Covid-19 on people experiencing poverty and vulnerability,* European Anti-poverty network report, July 2020.

[109] *Ibidem.*

[110] *Ibidem.*

[111] Sant'Egidio, *Dati e proposte per mettere fine alla "eterna zona rossa" per gli anziani ricoverati in strutture assistenziali*, 28 April 2021.

[112] Malgesini, G., *The impact of Covid-19 on people experiencing poverty and vulnerability,* cit.

[113] *Ibidem.*

[114] *Ibidem.*

[115] *Ibidem.*

31,2%, 30%, 27,3% respectively), while Czechia, Slovenia, Finland and Denmark the lowest ones (12,5%, 14,4%, 15,6%, 16,3% respectively)[116]. This percentages raised after the pandemic outbreak, as revealed by the first analysis: in Italy for instance, people that were working in the touristic sector or did not have regular jobs were among those asking for help. In general, a 30% increase in the food parcels was recorded[117]. Eurostat then revealed that the percentage of people at risk of poverty or social exclusion raised by 4,47% in 2020. Of this 96 million, 27 million are severely materially and socially deprived[118].

On the other hand, even though some studies claim that lockdown measures are likely to have worsened social cohesion in Europe[119], ONGs claim that the number of volunteers raised compared to before the disease's outbreak[120]. It must be said, however, that according to the European Parliament, especially in the first part of the pandemic, episodes of discrimination, against Asian and Italians were reported[121]; moreover, border closures were also invoked as a solution against the contagion and asylum seekers were accused of spreading the virus[122]. This was also evident in the web, where xenophobic messages were among the most shared ones[123].

Scapegoating is also consistent with the response of a population already stressed by the local lockdown; indeed, hate speech and mistrust towards the authority is also triggered by the decision to impose quarantines and general closures[124]. The pandemic of Covid-19, indeed, is changing our lifestyle and habits, from physical activity, to eating habits[125].

Social distancing measures also imposed school closures: even though the effectiveness of this policy on the virus transmission is strongly debated, almost 138 countries in the world closed schools within their territory, thus widening the gap between children with lower-income families and higher-income families; online

---

[116] Eurostat, *Over 20% of EU population at risk of poverty or social exclusion in 2019*, 16th of October 2020.

[117] La Repubblica, *Sant'Egidio: "Alla mensa +30% di utenti, raddoppiati i pasti da asporto. Arrivano anche nuovi poveri*, 29th of May 2020.

[118] Eurostat , *One in five people in the EU at risk of poverty or social exclusion*, 15th of October 2021

[119] University of Oxford, *Poverty and inequality surge across Europe in the wake of COVID-19*, 29 October 2020.

[120] Mira, A., *Pandemia sociale. Effetto Covid e nuove emergenze: a Roma la povertà mangia le famiglie*, Avvenire, 23 July 2020.

[121] European Parliament, *European Parliament resolution of 17 April 2020 on EU coordinated action to combat the COVID-19 pandemic and its consequences* [2020/2616(RSP)], 17th of April 2020.

[122] Devakumar, D., et al., *Racism and discrimination in COVID-19 responses*, The Lancet, 395, 10231, 1194, 2020.

[123] Shild, L., *"Go eat a bat, Chang!": An Early Look on the Emergence of Sinophobic Behavior on Web Communities in the Face of COVID-19*, arXiv preprint arXiv, 2004, 04046, 2020.

[124] Gaidhane, S., et al., *Depression, anxiety and stress among the general population in the time of COVID-19 lockdown: A cross-sectional study protocol*, International Journal of Research in Pharmaceutical Sciences, 11, 1, 2020.

[125] Cancello, R.; Soranna, D.; Zambra, G.; Zambon, A.; Invitti, C. *Determinants of the Lifestyle Changes during COVID-19 Pandemic in the Residents of Northern Italy*, Int. J. Environ. Res. Public Health, 17, 6287, 2020; Martinez, V., et al., The assessment of lifestyle changes during the COVID-19 pandemic using a multidimensional scale, Revista de Psiquiatría y Salud Mental, 14, 1, 16-26, 2021.

learning, indeed, is more difficult in low-income households for a series of reasons.[126] In Europe, 7% of the children do not have access to internet, 5% do not have a proper place where to do homework, and 5% do not have access to books at the appropriate reading level[127]. Furthermore, schools are also important for alimentary education and food security: according to Eurostat, indeed, 6,6 % of households with children in the EU cannot afford a "meal with meat, fish, or a vegetarian equivalent every second day"[128]. Moreover, families which were already relying on social assistance or school support (but even those who were not in need) were at particular risk during local lockdowns[129]. In the long run, also given the economic situation poor households are already living, this may affect the children's health, wellbeing, and future learning decisions[130]. This is extremely problematic, given the statistics on early leavers from education (18-24 years) recorded before the pandemic: in 2019, indeed, 11,9% of European young men and 8,4% of European young women left the educational path[131]. The lowest numbers were recorded in Croatia (2,2%), Greece (3,8%) and Slovenia (4,1%), while the highest ones are documented in Malta (16,7%), Spain (16%), and Romania (15,6%)[132].

Social consequences of Covid-19 also include an increase in the gender gap; according to the *"2021 report on gender inequality in the EU"*, the pandemic was a major challenge for the gender equality, exacerbating existing inequalities in "almost all areas of life"[133]; for instance, women obtained half as many jobs as men, women's share of unpaid work increased, and women had to cope with children care more than fathers while teleworking[134]. Moreover, pandemic also triggered a rise in domestic violence against women[135].

Clearly, the pandemic will also have demographic consequences on the European societies: Europe, indeed, is still experiencing the aftermath of the 2008 financial crisis and this is more evident in Southern Europe than in the other countries of the Union[136]. Here, uncertainty for the future is associated to the lowest fertility rates of the entire EU; moreover, in this sense, the Covid-19 containment strategies will also impact countries such as Spain and Italy, where the grandparental childcare is more common[137].

---

[126] Van Lancker, W., Parolin, Z., *COVID-19, school closures, and child poverty: a social crisis in the making,* The Lancet, 5, 5, 243-244, 2020.

[127] *Ibidem.*

[128] *Ibidem.*

[129] Malgesini, G., *The impact of Covid-19 on people experiencing poverty and vulnerability,* cit.

[130] Van Lancker, W., Parolin, Z., *COVID-19, school closures, and child poverty: a social crisis in the making,* cit.

[131] Eurostat, *Early leavers from education and training,* April 2020.

[132] Eurostat, *Early leavers from education and training by sex and labour status*, dataset, April 2021.

[133] European Union, *2021 report on gender inequality in the EU,* 2021.

[134] European Institute for gender inequality, *Covid-19 derails gender equality gains*, 05 March 2021

[135] *Ibidem.*

[136] Luppi, F., Arpino, B., Rosina, A., *The impact of COVID-19 on fertility plans in Italy, Germany, France, Spain, and the United Kingdom,* Demographic Research, 43, 47, 1399-1412, 2020.

[137] *Ibidem.*

On the other hand, it is a fact that Europe in general has an ageing population and low birth rates, but is also experiencing "depopulation of rural areas, internal and external migration and brain drain"[138]: thus, the current situation may have further impact on the less developed regions (already facing high unemployment) and may also affect healthcare services[139].

From the political point of view, Covid-19 had effects on democracy too: indeed, a huge debate is still going on regarding the declaration of "state of emergency" and the adoption of "emergency powers" by the states[140]. Importantly, the declaration of "state of emergency" allows a government to increase its powers, but also diminish the ones of the Judiciary and of the Parliament, negatively affecting the system of checks and balances as well as restricting human rights and fundamental freedoms[141]. Thus, the fear of citizens that these instruments may lead to an authoritarian decay, as also happened in the past[142]. However, constitutions normally foresee guarantees for the democracies to survive these periods of crisis: Parliaments must be informed of the declaration of the state of emergency, which has an expected end, they must vote *ex ante* or *ex post* the act or decrees approved under the state of emergency, and government's activity is supervised by MPs[143].

However, among the countries of the European Union, in two cases, the adoption state of emergency, usually implemented unanimously or with the involvement of the oppositions, did not foresee guarantees for the opposition; the European Parliament, indeed:

> "deems it totally incompatible with European values both the decision from the Hungarian Government to prolong the state of emergency indefinitely, to authorise the Government to rule by decree without time limit, and to weaken the emergency oversight of the Parliament, and the steps taken by the Polish Government namely changing the electoral code against the judgment of Constitutional Tribunal and provisions laid by law to hold Presidential elections in the middle of a pandemic, which may endanger the lives of Polish citizens and undermine the concept of free, equal, direct and secret elections as enshrined in the Polish Constitution"[144].

Actually, Poland and Hungary were already accused of illiberal backslidings long before the pandemic and the European Commission had initiated the Article 7 procedure against those two countries in 2017 and 2018, respectively[145]. How-

---

[138] The EU's Assembly of Regional and Local Representatives, *The Covid-19 crisis can deepen Europe's demographic decline, regions and cities warn*, 15th of October 2020.

[139] *Ibidem*

[140] Marzocchi, O., *The impact of Covid-19 measures on Democracy, the Rule of Law and Fundamental rights, Policy Department for Citizens' Rights and Constitutional Affairs,* Briefing requested by the LIBE committee, Monitoring Group on Democracy, Rule of Law, Fundamental Rights, 2020.

[141] *Ibidem.*

[142] *Ibidem.*

[143] *Ibidem.*

[144] European Parliament, *European Parliament resolution of 17 April 2020 on EU coordinated action to combat the COVID-19 pandemic and its consequences,* cit.

[145] Michelot, M., *The "article 7" proceedings against Poland and Hungary: what concrete effects?* Institut Delors, 2019.

ever, as noted by the European Parliament on the 16th of January 2020, the process "resulted in any significant progress"[146]. Moreover, the two countries belong to the so called "Visegrad group", which also includes Czech Republic and Slovakia: all four countries, at present date, are being ruled by populist parties[147] and all of them were facing "illiberal swerving" before the pandemic[148]. Thus, the current situation offers several opportunities to strengthen the power, as it undermines accountability by strengthening the role of the experts, it offers a great opportunity to fan the flames of the increasing social unrest and it gives leaders the possibility to weaken the civil society, the opposition, and the powers of Parliaments[149]. According to PhDr Guasti, while democracy in Czech Republic and Slovakia proved to be resilient to the shock, it is not the same for Poland and Hungary. In the Baltic state, indeed, according to Guasti, the ruling Party tried to undermine the democratic rule by forcing the general vote via mail[150]. The problem laid in the fact that citizens were asked to deliver their data via mail to the Polish Post, thus violating the General Data Protection Regulation, in the opinion of Guasti. OSCE too took position against this practice, which could constitute a breach of the principles of legality[151].

More worrying is the situation in Hungary, where the Government is accused of having transformed Hungary in an authoritarian country: in March 2020, indeed, the Parliament passed a law adopting the state of emergency without a definite end[152]. No elections or referenda can take place during the state of emergency and the Prime Minister can rule by decree; moreover, according to a new law, those who "disseminate misleading information risk five years of jail"[153].

In addition, suspensions of the rule of law have been recorded in North Macedonia and Serbia too, and violations of the privacy were documented in Montenegro and Bosnia-Herzegovina, along with accuses of intimidations by the authorities. However, both the latter countries already had a weak rule of law: the pandemic only sped up an ongoing process[154]. We can probably state that the same considerations apply both for Hungary and Poland.

Among the group of Visegrad, Slovakian Prime Minister Matovic had to resign due to a scandal related to the Russian vaccine *Sputnik:* Matovic, indeed, decided to order 2 million doses of the serum without consulting his coalition partners. Importantly, the said vaccine has not yet been approved by the European authorities[155].

---

[146] European Parliament, *European Parliament resolution of 16 January 2020 on ongoing hearings under Article 7(1) of the TEU regarding Poland and Hungary (2020/2513(RSP)),* 16 January 2020.

[147] Guasti, P., *The Impact of the COVID-19 Pandemic in Central and Eastern Europe,* Democratic Theory, 7, 2, 47-60, 2020.

[148] Bustikova, L., Guasti, P., *The Illiberal Turn or Swerve in Central Europe?* Politics and Governance, 5, 4, 166–176, 2017.

[149] Guasti, P., *The Impact of the COVID-19 Pandemic in Central and Eastern Europe,* cit.

[150] *Ibidem.*

[151] *Ibidem.*

[152] *Ibidem.*

[153] *Ibidem.*

[154] Fruscione, G., *Covid-19 in the Balkans: The Virus of Authoritarianism,* ISPI, 30 April 2020.

[155] Muller, R., *New Slovak PM seeks to mollify Russia in vaccine row,* Reuters, 9 April 2021.

Hungary too bought and administered Sputnik's doses to its citizens[156], while Germany has recently said that it will start talks to secure supplies of the serum[157]. The race for the vaccine has thus been compared to the 50s' space race and is dividing the European states[158]: the Russian vaccine, indeed, has been defined a "soft power tool", since Russia also delivered it to Mexico, Philippines, India, South Korea, and Argentina, among others[159]. Further evidence of the current strategic importance of the vaccine is given by the tensions between the UK and the EU regarding the reciprocity of vaccine exports: the European Union, indeed, exported 10 million doses to the UK, but on the other hand, UK exported zero[160], further complicating UK-EU relationship, already strained by the post-Brexit Northern Ireland issue[161].

On the other hand, the vaccine issue is also related to the fake news issue: according to many analysts, in fact, Russia and China are engaged in a disinformation campaign pushing national media to public fake news in different languages and promote Chinese and Russian serums[162]. In fact, this was also noticed before the Pandemic, at least for Russia[163]. In June 2020, EU announced that it was taking action to stop disinformation[164], and decided to collaborate with the social networks to stop the spread of fake news[165]. Recently, Denmark and Italy made a declaration defending press freedom against interferences from China and Russia[166]. Moreover, according to the *Search for Common ground*'s study, financed by the EU, "inequitable access to pandemic-related information, confusion on what information sources and channels to trust, misinformation, and rumours determined mistrust towards the European governments, as well as rising inflation and prices of basic goods, loss of jobs, divisive or politicized rhetoric, heavy handed enforcement of pandemic protocols, inequities and disparities in access to pandemic-related support, and mismanagement of pandemic relief funds[167].

---

[156] Connolly, K., *How Russia's Covid vaccine is dividing Europe*, BBC, 17th of April 2021.

[157] Martuscelli, C., Gehrke, L., *Russia's Sputnik vaccine injects divisions into EU*, POLITICO, 14th of April 2021.

[158] *Ibidem*; Connolly, K., *Sputnik V: How Russia's Covid vaccine is dividing Europe*, cit.

[159] Connolly, K., *How Russia's Covid vaccine is dividing Europe*, cit.

[160] Sandle, P., Chalmers, J., *EU blames AstraZeneca as vaccine battle with UK deepens*, Reuters, 22nd of March 2021.

[161] Fleming, S., et al., *Tensions between EU and UK inflamed over Northern Ireland*, Financial Times, 3rd of March 2021.

[162] Emmott, R., *Russia, China sow disinformation to undermine trust in Western vaccines*: EU, Reuters, 29th of April 2021.

[163] Burki, T., *Vaccine misinformation and social media*, The Lancet, volume 1, 6, 258-259, 2019; Walter, D., et al., *Russian Twitter Accounts and the Partisan Polarization of Vaccine Discourse*, 2015-2017, AJPH open-themed research, 110, 5, 2020; Broniatowski, D., *Weaponized Health Communication: Twitter Bots and Russian Trolls Amplify the Vaccine Debate*, AJPH open-themed research, 108, 10, 2018.

[164] European Commission, *Coronavirus: EU strengthens action to tackle disinformation*, 10th of June 2020.

[165] European Commission, *Tackling coronavirus disinformation*, accessed on the 29th of April 2021.

[166] Marzocchi, O., *The impact of Covid-19 measures on Democracy, the Rule of Law and Fundamental rights*, cit.

[167] Search for Common Ground, *Trust in authorities – the golden ticket to successful COVID-19 vaccine roll-out in conflict settings*, January 2020.

Finally, as we are dealing with a pandemic, we must also include in our analysis the sanitary response of the European states as well as the health developments. First, we must recall that at the very beginning of the pandemic in Europe, some countries introduced limits on exports of protective medical equipment, as recently happened with vaccines[168]. Moreover, although other countries have learned important lessons from pandemics such as SARS, H1N1 and Ebola and were therefore very well prepared, the Covid-19 pandemic highlighted the lack of preparedness of the EU and several MSs, including those with usually efficient health systems[169]. In the aftermath of the SARS epidemy, EU states decided to create an agency dedicated to providing a pan-European response to a possible pandemic event, the European Centre for Disease Prevention and Control (ECDC), which has proven to work efficiently during the Zika and Ebola outbreaks[170]. However, shortcomings in cooperation between the Agency and the MSs have been documented; in addition, recommendations issued by the ECDC during the pandemic demonstrated the importance of a coordinated and binding approach, which was however lacking[171]. The ECDC, moreover, pays the price of the cuts on healthcare imposed by the financial crisis on the MSs; this was evident, for instance, in Italy, where health investments *per capita* declined until 2013 and only then began to slowly increase. However, EU decided to invest 350 million euros to support the development of vaccines and before Covid-19 650 million euros had been destined to vaccination research[172]. Moreover, the EU also invested 5,1 billion in the EU4Health Programme, the largest health programme ever in monetary terms: the funds will be available for MSs, Health organizations and NGOs and are aimed to strengthen the health systems in Europe[173]. Moreover, EU also announced the creation of a Recovery Plan, the "EU's largest stimulus package": 1,8 trillion euros that will be used to create a "greener, more digital and more resilient Europe"[174].

In conclusion, European countries were also among the most involved in the research on Covid-19: as shown by a study of Mei Hsiu-Ching Ho and John S. Liu, Italy and the UK were among the top five countries for quality and quantity of published research on the issue (the third and the fourth, respectively, while China and the US were the first two); France, Germany, Switzerland, Spain and the Netherlands were in the top 20 too[175]. The rapid growth of studies demonstrates the enormous efforts made to find solutions for the pandemic and the same studies have been extremely useful in developing vaccines and treatments to combat the disease[176].

---

[168] Anderson, M., *Covid-19 exposes weaknesses in European response to outbreaks*, BMJ, 368 1075, 2020.

[169] Renda, A., Castro, R., *Towards Stronger EU Governance of Health Threats*, European Journal of Risk Regulation, 11, 2, 273-282, 2020.

[170] *Ibidem.*

[171] *Ibidem.*

[172] European Commission, *EU support for vaccines*, accessed on the 29th of April 2021.

[173] European Commission, *EU4Health 2021-2027 – a vision for a healthier European Union*, accessed on the 29th of April 2021.

[174] European Commission, *Recovery plan for Europe*, accessed on the 29th of April 2021.

[175] Hsiu-Ching Ho, M., Liu, J., *The swift knowledge development path of COVID-19 research: the first 150 days*, Scientometrics, 126, 2391–2399, 2021.

[176] *Ibidem.*

2.2 Europe's 21st century Migration Period

2.2.1 The European routes

As pointed out by Bozorghmer et al., Europe is currently facing several crises: the Covid-19 pandemic, indeed, "converges and interacts" with the migrant crisis[177]. In fact, the pandemic did not stop the migratory fluxes, but MSs nonetheless suspended asylum procedures, transfers to other countries in accordance with the Dublin Agreement and returns[178]. Moreover, the European Parliament claimed that refugees arriving in the Greek islands do not have access to health care and are thus in potential risk of contracting Covid-19[179]. This situation, however, is not new: in fact, the so called "refugee crisis" is an ongoing process that started in 2011, during the so-called *Arab springs*[180], and continued throughout the following years, during which migrants increasingly attempted to cross the Mediterranean[181]. In ten years more than 5.634.600 asylum applications were documented in the EU, excluding the second time applications[182]. However, while we primarily associate the crisis to the Mediterranean shipwrecks, there are other migratory routes which end up in the European Union on which we will now focus. Importantly, we will focus predominantly on the main reasons why migrants arrive in Europe, rather than on the European policies of admission or on the countries of settlements: our intention, in fact, is to show that these migrations must be seen as historical phenomena rather than isolated mass arrivals.

The Eastern Borders route

The Eastern Borders route is the least important of the routes leading to the European borders, and mainly involves people not associated with migration, but with smuggling of goods; however, Vietnamese, along with the Afghans, Iranians, Iraqis, and Bengalese, are among the main populations usually trying to enter Europe via the Latvian-Russian border[183]. More Afghans, on the other hand, have tried to cross the Hungarian-Ukrainian border; so far, nearly 14.000 people have attempted to cross[184].

The Albanian circular route

Soon after the end of the Communist rule in Albania in 1991, 300.000 Albanians left their country, and the trend continued in the following years: in 1995, 295.000

[177] Bozorghmer, K., et al., *COVID and the convergence of three crises in Europe,* The Lancet, 5, 5, 247-248, 2020.

[178] Marzocchi, O., *The impact of Covid-19 measures on Democracy, the Rule of Law and Fundamental rights,* cit.

[179] *Ibidem.*

[180] Lendaro, A., *A 'European Migrant Crisis'? Some Thoughts on Mediterranean Borders*, Studies in ethnicity and Nationalism, 16, 1, 148-157, 2016

[181] Eurostat, *Asylum applications (non-EU) in the EU Member States, 2008-2020*, 27 April 2021, accessed on 30/04/2021.

[182] *Ibidem.*

[183] Frontex, *Migratory Routes, Eastern Border Route,* accessed on 30/04/2021.

[184] *Ibidem.*

people left Albania, mainly for Greece, and the following year the number raised to 428.000[185]. Then in the period between 2000 and 2008 Albanians continued to migrate but in smaller numbers (40.000 in the first decade of the century)[186]. Importantly, Albanians were generally employed in seasonal works at the end of which they returned home, hence the name "circular route"; however, subsequent regularization programs resulted in Albanians settling in Greece.[187] Nonetheless, in 2009 irregular border crossing began to increase again: in that year and in 2010 more than 75.500 Albanians entered the EU (in Greece)[188]. In 2010, however, the number dropped sharply to 5269, and between 2012 and 2019, "only" 50.014 Albanians entered Greece[189].

The Eastern Mediterranean route

In addition to migrants coming from Albania, Greece, the "eastern gate of Europe", has also seen arrivals from Turkey; migrants usually arrive either by sea or via land[190]. However, these two routes are complementary, since statistics show that once numbers increase in one, they decrease in the other[191]. Furthermore, as we will see for Italy, the geopolitical changes in the neighbouring regions (north Africa and Middle East) are reflected in the migrants' arrivals: for instance, until 2011, Syrians were no more than 6% of the asylum seekers, being Afghans and Pakistanis the most represented[192]. Then, the situation changed in 2012, when Syrians became the third nationality per arrivals (12,2%); the following year they were the 37% of the newcomers (see *Table 1*). In 2013 Syrians also arrived in Italy, where they became the first nationality per arrival (26,3%) (*see Table 2*).

Table 1: *Arrivals in Greece* (Data from UNHCR and Greek Ministry for the Protection of the Citizen, consulted in May 2021)

YEAR	TOTAL NUMBER OF ARRIVALS	FIRST COUNTRY OF ORIGIN		SECOND COUNTRY OF ORIGIN		THIRD COUNTRY OF ORIGIN	
2009[193]	85.626	Afghanistan	20,8%	Palestine	12,5%	Somalia	9%

[185] Maroukis, T., & Gemi, E. *Albanian circular migration in Greece: beyond the state?* In A. Triandafyllidou (Ed.), Circular Migration between Europe and its Neighborhood: Choice or Necessity? Oxford University Press, 68-89, 2013.

[186] *Ibidem*; Frontex, *Number of illegal border-crossings to the European Union (EU) over the circular route from Albania to Greece from 2009 to 2019,* 28 April 2020, In Statista. Retrieved April 30, 2021.

[187] Maroukis, T., & Gemi, E. *Albanian circular migration in Greece: beyond the state?* cit.

[188] Frontex, *Number of illegal border-crossings to the European Union (EU) over the circular route from Albania to Greece from 2009 to 2019,* cit.

[189] *Ibidem.*

[190] Triandafyllidou, A., *Migration in Greece Recent Developments in 2014,* Hellenic foundation for European and Foreign Policy, Report prepared for the OECD Network of International Migration Experts Paris, 6-8 October 2014, European University Institute, Robert Schuman Centre for Advanced Studies, 2014.

[191] *Ibidem.*

[192] Data from the Greek Ministry for the Protection of the Citizen.

[193] Data refer to apprehensions, not to migrants.

2010[194]	96.956	Afghanistan	29,1%	Pakistan	9,1%	Palestine	7,7%
2011[195]	86.622	Afghanistan	32,9%	Pakistan	23%	Bangladesh	6%
2012[196]	64.783	Afghanistan	25,5%	Pakistan	17,1%	Syria	12,2%
2013	22.962	Syria	37%	Afghanistan	27,9%	Pakistan	17,3%
2014	45.421	Syria	38,2%	Afghanistan	13,6%	Pakistan	4%
2015	861.630	Syria	56,1%	Afghanistan	24,4%	Iraq	10,3%
2016	177.234	Syria	47%	Afghanistan	24%	Iraq	15%
2017[197]	29.718	Syria	41,3%	Iraq	19,5%	Afghanistan	11,4%
2018	50.508	Afghanistan	28%	Syria	24%	Iraq	18%
2019	74.613	Afghanistan	38%	Syria	23%	Iraq	6%
2020	15.696	Afghanistan	35,2%	Syria	22,7%	Dem.Rep. Congo	10,3%
TOT	**1.611.769**						

Despite the relative proximity of Greek islands and the Turkish coast, many migrants continue to drown in the Aegean Sea; unfortunately, deaths on the land route are also documented[198].

According to UNHCR 11.400 refugees and migrants reached the Greek islands in 2013 (only 3600 arrived in the previous years): as said, most of them were Syrians fleeing the war and heading the islands of Lesvos, Chios, Samos, Kos and Leros[199]. However, until 2015, the primary destination of those crossing the Mediterranean was mainly Italy. In 2014, 170.000 people arrived on the Italian shores; meanwhile, "only" 43.500 were entering Greece via sea[200]. Then, the situation changed the following year: 856.723 arrivals by sea and 4.907 by land were recorded in Greece[201], while in the same period, Italy documented 153.842 new migrants

[194] *Ibidem.*

[195] *Ibidem.*

[196] *Ibidem.*

[197] Data are only referred to the arrivals by sea.

[198] Trakilović, M., *"On this path to Europe" – the symbolic role of the 'Balkan corridor' in the European migration debate* in Cultures, Citizenship and Human Rights, edited by Rosemarie Buikema, Antoine Buyse and Antonius C.G.M. Robben, Routledge, 2020.

[199] *Ibidem*

[200] UNHCR, *The sea route to Europe: Mediterranean passage in the age of refugees*, 2015.

[201] UNHCR, *Operation Portal, Refugee situations, Mediterranean Situation, Greece,* updated to the 25th of April 2021, accessed on 01/05/2021.

landed on its shores[202]. In 2016 numbers were almost the same, as Italy recorded 181.436 asylum seekers (and 4578 deaths) and Greece 173.450 sea arrivals plus 3784 land arrivals. Then, in 2017, Italy again became the first destination of migrants crossing the sea: 119.369 asylum seekers were reported, compared to the 36.310 ones of Greece (sea and land arrivals)[203].

The evolution of the numbers of migrant arrivals must be linked to the agreement made by the EU with Turkey and also connected to the existence of the *Balkan corridor,* on which we will focus later. Indeed, this *corridor* is sometimes described as a "short living phenomenon" since numbers of refugees remained high for two years (2015-2016), following which the said agreement was signed[204]. Indeed, while 1.082.182 documented arrivals were recorded in the period 2014-2016 (98% sea arrivals), this number dropped to 177.127 in the period 2017-2020 (74% sea arrivals)[205]. According to the agreement, irregular migrants crossing from Turkey to Greece will be returned to Turkey; however, "for every Syrian being returned to Turkey from the Greek islands, another Syrian will be resettled to the EU"[206]. Moreover, EU allocated 6 billion euros to Turkey to support the 3,7 million refugees in the country with education, healthcare, and Covid-19 assistance[207]. However, this agreement has been widely criticized and many accused Europe of dismissing its humanitarian responsibilities; moreover, Turkey's government has been accused by Greece of using asylum seekers as leverage to obtain additional money and political concessions from Europe[208]. Indeed, despite the agreement, Turkey let refugees flow into Greece in March 2020 as a way to express its discontent with Europe[209].

In the period between 2009 and 2020, 1.611.769 asylum seekers arrived in Greece, both by land and sea; the main countries of origin of migrants are Syria, Afghanistan, Iraq, Pakistan and Democratic Republic of Congo.

The Balkan route or the *Balkan corridor*

The Western Balkan route is a migratory route which crosses the Balkans (Albania, Bosnia-Herzegovina, Croatia, Montenegro, Serbia and North Macedonia); it starts in Greece, where refugees and asylum seekers arrive from Turkey by land or sea, and it continues up to the borders of Europe[210]. An alternative route runs from

---

[202] UNHCR, *Operation Portal, Refugee situations, Mediterranean Situation, Italy,* updated to the 25th of April 2021, accessed on 01/05/2021.

[203] UNHCR, *Operation Portal, Refugee situations, Mediterranean Situation, Greece*, cit., UNHCR, *Operation Portal, Refugee situations, Mediterranean Situation, Italy,* cit.

[204] Trakilović, M., *"On this path to Europe" – the symbolic role of the 'Balkan corridor' in the European migration debate* in Cultures, cit.

[205] UNHCR, *Operation portal, Refugee situation, Mediterranean situation, Greece,* cit.

[206] European Parliament, *Legislative train 03.2021 - 8 towards A New Policy on Migration*, 2019.

[207] European Commission, *EU signs final contracts under the €6 billion budget of the Facility for Refugees in Turkey*,17th of December 2020.

[208] Terry, K., *The EU-Turkey Deal, Five Years On: A Frayed and Controversial but Enduring Blueprint,* The online journal of the Migration Policy Institute, 8 April 2021.

[209] *Ibidem.*

[210] Idemudia, E., *Travelling Routes to Europe,* in Idemudia, E., Boehnke, K., *Psychosocial Experiences of African Migrants in Six European Countries*, Social indicators Research Series, Springer, 33-49, 2020.

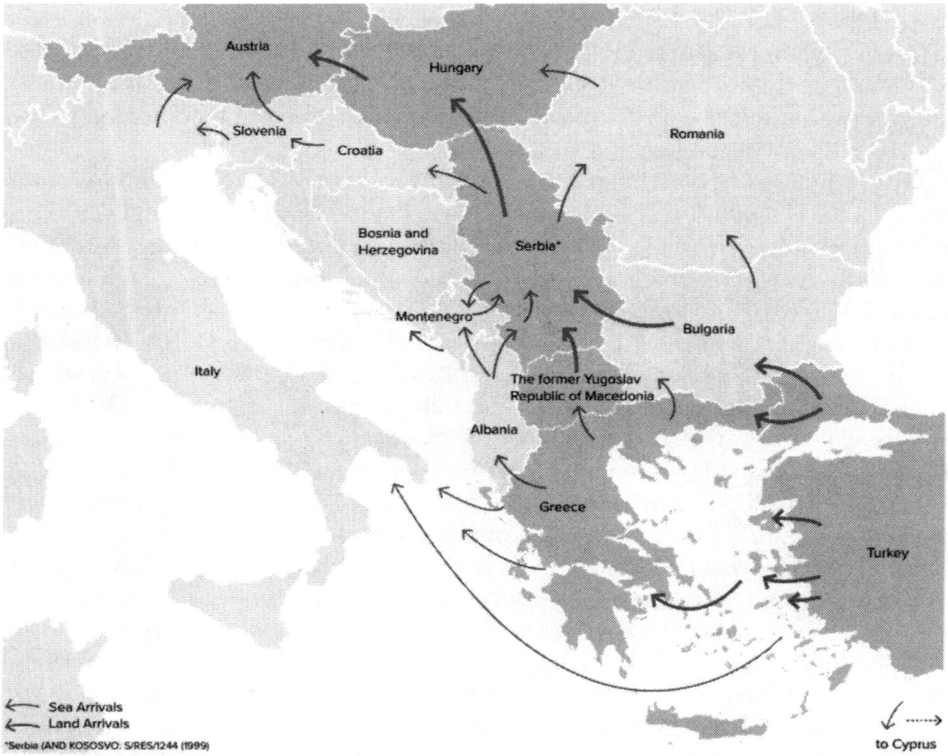

Figure 13: Balkan route (UNHCR).

Serbia to Bulgaria to the other EU MSs[211]. Once at the EU borders, migrants' destinations are mainly Hungary, Austria, Germany and Sweden[212]. Asylum seekers' countries of origin are mostly Syria, Afghanistan, Iran[213], Morocco and Pakistan[214].

The existence of the *Balkan Corridor* went unnoticed until 2015, when Austrian authorities discovered a truck with seventy-one bodies of migrants; then, when 350.000 migrants marched through Hungary towards the border with Austria, the Hungarian Government decided to close the borders with Croatia and Serbia, thus re-routing the *corridor*[215]. In August 2015 Germany decided to open the borders to the refugees[216] and in November border crossing were allowed for Syrians, Afghans and Iraqis[217].

---

[211] *Ibidem.*

[212] *Ibidem.*

[213] Frontex, *Western Balkan route,* accessed on 30/04/2021.

[214] Roksandic, *S., et al., Migration in the Time of COVID-19—Policy Responses and Practices in Croatia Concerning the Western Balkan Routes and Readiness for the Post-COVID-19 Society in Which the Right to Health Care for the Most Vulnerable Is Guaranteed*, Front. Hum. Dyn., 2021.

[215] Abikova, J., Piotrowicz, W., *Shaping the Balkan corridor: Development and changes in the migration route 2015-16*, International Migration, IOM, Wiley, 59,2021.

[216] *Ibidem.*

[217] Momin, S. *Human Rights Based Approach to Refugees: Look at the Syrian Refugee Crisis and the Responses from Germany and the United States*, Duke Forum for Law and Social Change, 9, 55-80, 2017.

In 2016 new rules were applied in Austria, Germany and Slovenia and Afghanis were no longer allowed to enter these countries, while Macedonia built a fence at the border with Greece[218]. Moreover, in March, the Balkan countries announced the closure of their borders to irregular migration, thus putting the migrants which had already had already started the journey in an indeterminate state[219]. There are no official statistics on the migratory movements in the Balkans: the EU Parliament estimates that almost 600.000 people transited along the *corridor* in 2015, while the UNHCR increases the number to 815.000.

Despite being entirely overland, this route is dangerous for a number of reasons: first of all, when winter arrives, migrants have to face the Balkan cold; secondly, migrants frequently report physical and sexual violence, torture, and extortions by traffickers; thirdly, the path is not simple, and several migrants lost their lives when trying to cross the Danube at the Romanian-Bulgarian border; finally, pushbacks and use of violence by Hungarian state authorities have also been documented[220].

To date, the pandemic of Covid-19 is making things more difficult for migrants crossing the Balkans; this is not only because countries have closed the borders, but also because of the sanitary conditions in refugee centres, as reported by several NGOs and independent media[221]; in addition, recently, MEPs which were visiting the border between Bosnia and Croatia were prevented from carrying their inspection[222].

The *Atlantic route*

Among the possible routes, West Africans migrants also choose the Western Mediterranean one, known as the *Atlantic route*: in recent months, the popularity of this migratory path has increased, due to a series of reasons that we will discuss later[223]. People from Senegal, Gambia, Mauretania, Western Sahara and Morocco, leave for the Canary Island, sometimes from their own countries but mostly from Western Sahara or Morocco: this is a trip of several hundreds of kilometres, which is made onboard of makeshift boats or pirogues[224] . In 2020 only, 23.000 people arrived in the Spanish islands, but we cannot estimate the number of migrants who left their homes since sometimes they are intercepted or rescued at sea, while sometimes they drown: for instance, at least 500 people lost their lives on this

---

[218] Šantić, D., *The Balkan Migration Route: Reflections from a Serbian Observatory* in Towards understanding of Contemporary Migration Causes, Consequences, Policies, Reflections, edited by Mirjana Bobić, Stefan Janković, Institute for Sociological Research, Belgrade, 221-241, 2017.

[219] *Ibidem.*

[220] UNHCR, *Desperate journeys, Refugees and migrants entering and crossing Europe via the Mediterranean and Western Balkans routes,* 2017.

[221] Roksandic, *S., et al., Migration in the Time of COVID-19 — Policy Responses and Practices in Croatia Concerning the Western Balkan Routes and Readiness for the Post-COVID-19 Society in Which the Right to Health Care for the Most Vulnerable Is Guaranteed,* cit.

[222] *Ibidem.*

[223] Affeldt, J., Seaman, M., A Gateway Re-opens: the growing popularity of the Atlantic route, as told by those who risk it, MMC Research Report, February 2021.

[224] *Ibidem.*

route between January and November 2020[225]. In 2016, 675 migrants left Africa for the Canary Island; in 2017, they were 424; in 2018, 1307 (69 ship arrivals, and 18 recorded shipwrecks), in 2019, 2698 (133 ship arrivals and 26 recorded shipwrecks); in 2020 the number increased to 23.023 (545 ship arrivals and 41 documented shipwrecks)[226].

The route was started in 2006, when 31.678 migrants arrived in the Spanish ar chipelago onboard of cayucos, small boats; Spain[227] had already signed an agreement for repatriation with Mauretania and in 2006 another one was signed with Senegal, while the Frontex operations (*European Border and Coast Guard Agency*) extended its activity there too[228]. The Spanish *Guardia civil* is also present in Mauretania and Spain to contrast activities of human trafficking[229].

The Western Mediterranean route

Already in 2005 thousands of sub-Saharan Africans tried to overstep the fences dividing the Spanish enclave of Melilla and Morocco; in the previous decade, the route was mainly used by Moroccans and Algerians trying to enter Spain, but this changed at the end of the 90s[230]. Currently, the route is used by Western Africans and Maghrebians (mainly Moroccans and Algerians) to enter Europe[231].

Migrants leave Africa from Tangier, Morocco (90%)[232], or from Algeria (9%)[233] and head to Tarifa, southernmost city in Spain, with the help of smugglers, but they also try to cross the border of Ceuta and Melilla, Spain's two enclaves, either through false Moroccan passport or by climbing the fences with makeshift ladders, after waiting in the neighbouring forests for months[234]. When crossing the Strait of Gibraltar, migrants are transported on "mother boats" into Spanish water, from where they continue the trip via dinghies or makeshift boats[235]. In alternative, they sail from the Atlantic coast but instead of heading the Canary island, they try to reach Spain: smugglers, indeed, often choose areas in which the control of the authorities is looser[236].

---

[225] *Ibidem.*

[226] IOM, *Irregular migration towards Europe, Western African Route: Migration to the Canary Island,* 18 November 2020.

[227] UNHCR, Operational Portal, Refugee Situation, Mediterranean Situation, *Spain*, accessed on the 30th of April 2021.

[228] Affeldt, J., Seaman, M., *A Gateway Re-opens: the growing popularity of the Atlantic route, as told by those who risk it,* cit.

[229] *Ibidem.*

[230] Idemudia, E., Travelling Routes to Europe, cit.

[231] *Ibidem.*

[232] Fernandez-Sebastian M., *Maritime Dimension in the Fight Against Illegal Migration on the Western Mediterranean Route,* in: Ramírez, J., Biziewski J. (eds) Security and Defence in Europe. Advanced Sciences and Technologies for Security Applications. Springer, Cham, 2020.

[233] *Ibidem.*

[234] Idemudia, E., *Travelling Routes to Europe*, cit.

[235] Fernandez-Sebastian M., *Maritime Dimension in the Fight Against Illegal Migration on the Western Mediterranean Route,* cit.

[236] *Ibidem.*

Along this route, migrants' main countries of origin are Morocco, Guinea, Mali and Algeria, but also Côte d'Ivoire and Gambia (see Table 2).

Table 2

Year	Total number of arrivals	First country of origin		Second country of origin		Third country of origin	
2015	16.936	Syria	44,2%	Guinea (Conakry)	13,6%	Algeria	10,6%
2016	14.605	Algeria	21%	Côte d'Ivoire	21%	Gambia	12%
2017	28.449	Morocco	15,8%	Algeria	14,7%	Guinea (Conakry)	11,5%
2018	65.383	Guinea	20%	Morocco	20%	Mali	16%
2019	32.513	Morocco	25%	Algeria	16%	Guinea	15%
2020	41.861	Algeria	39,5%	Morocco	20,3%	Mali	12,6%
TOT	**199.747**						

The Central Mediterranean route

Italians use to remember the arrival of the *Vlora* ship carrying 20.000 migrants from Albania as the beginning of a new era: Italy was becoming a place of arrival, after years of emigration[237]. The years following the fall of the Berlin wall, in fact, were crucial for a series of reasons: the end of the communist rule in Eastern Europe, the associated mass migrations from beyond the former *Iron curtain* and the diversifications of the arrivals from Asia and Africa[238]. However, in the 90s and in the first decade of 2000, Italy saw a limited number of migrant arrivals on its southern shores, compared to those that would follow from 2011 onwards: this was also due to the policy of rejection of ejecting migrant boats, which was possible thanks to the 2008's agreement with Libya which has been widely criticized by NGOs and international observers such as *Human Rights Watch*[239], as well as by the ECHR[240]. According to these voices, indeed, the refoulment of migrants to Libya was contrary to the principle of *non-refoulment,* and endangered the lives of migrants, as reports from the North African country provided clear evidence of physical abuse, sexual violence in detention centres, as well as extortion and collusion with smugglers[241]. However, geopolitical transformations in North Africa have completely changed the scenario. While in 2010

[237] Il Post, *I primi sbarchi di massa degli albanesi, trent'anni fa,* 7th of March 2021.

[238] Colucci, M., *Storia dell'immigrazione straniera in Italia, dal 1945 ai giorni nostri,* Carocci Editore, Roma 2017.

[239] HRW, *Pushed Back, Pushed Around,* 21st of September 2009.

[240] Colucci, M., *Storia dell'immigrazione straniera in Italia,* cit.

[241] HRW, *Pushed Back, Pushed Around,* cit.

only 4500 people arrived, in 2011 the number rose to 62700 (see table 3). In the following years the numbers increased again, and Italy encountered several difficulties in managing the situation: moreover, many migrants fled from reception centres in Italy because they wanted to reach Northern European countries[242].

Nonetheless, as we can see from Table 3, from 2017 onwards the number of arrivals decreased sharply: to understand why, we must remember that in 2016, the EU-Turkey agreement was signed, as already mentioned. A similar agreement was signed between Italy and Libya: according to the Memorandum of Understanding signed in February 2017, Italy commits to give support to Libya in combating the irregular immigration[243], while both parties agree to strengthen the control over the Niger border[244]. Like the EU-Turkey agreement, the Libya-Italy agreement was also criticized because it did not mention the "legal migration", i.e. the right to migrate of people holding "refugee"status: this means that Italy was "outsourcing" the asylum management to Libya, which does not recognize the right to asylum or cannot process international protection claims[245]. Moreover, doubts were also expressed about the "temporary reception centres under the exclusive control of the Libyan Ministry of Interiors"[246]. In addition, the pandemic of Covid-19 also reduced the number of arrivals, although not in all the countries of departure[247]; we will focus on this issue later.

The "Central Mediterranean Route" is considered the most dangerous of the three routes, but it is nonetheless commonly used; the departure points used to be in Tunisia, Algeria and Morocco, but departures from Libya increased after the fall of Gaddafi regime[248]. The Libyan *Jamahiriya* in fact, maintained tight control over the coast but its demise was followed by a general instability, with the consequent growth of uncontrolled smuggling networks[249]. Recently, however, Tunisia has replaced Libya as the country with the highest number of migrant departures; on the other hand, the programmes enacted by the IOM, the UN and the African Union to evacuate migrants in Libya's detention centres are reducing pressure on the migratory route[250]. Another

---

[242] UNHCR, *The sea route to Europe: The Mediterranean passage in the age of refugees,* cit.

[243] "La parte italiana si impegna a fornire supporto tecnico e tecnologico agli organismi libici incaricati della lotta contro l'immigrazione clandestina, e che sono rappresentati dalla guardia di frontiera e dalla guardia costiera del Ministero della Difesa, e dagli organi e dipartimenti competenti presso il Ministero dell'Interno" (Memorandum d'intesa sulla cooperazione nel campo dello sviluppo, del contrasto all'immigrazione illegale, al traffico di esseri umani, al contrabbando e sul rafforzamento della sicurezza delle frontiere tra lo Stato della Libia e la Repubblica Italiana).

[244] "Le parti si impegnano a prendere azioni nei seguenti settori: 1) completamento del sistema di controllo dei confini terrestri del sud della Libia, secondo quanto previsto dall'articolo 19 del Trattato summenzionato" (Memorandum d'intesa sulla cooperazione nel campo dello sviluppo, del contrasto all'immigrazione illegale, al traffico di esseri umani, al contrabbando e sul rafforzamento della sicurezza delle frontiere tra lo Stato della Libia e la Repubblica Italiana).

[245] Palm, A., *The Italy-Libya Memorandum of Understanding: The baseline of a policy approach aimed at closing all doors to Europe?* EU Immigration and Asylum Law and Policy Droit et Politique de l'Immigration et de l'Asile de l'UE, 02 October 2017.

[246] *Ibidem.*

[247] Villa, M., Fact Checking: migrazioni (e Covid-19), ISPI online, 29th of July 2020.

[248] Idemudia, E., *Travelling Routes to Europe,* cit.

[249] *Ibidem.*

[250] *Ibidem.*

factor is the increasing danger of the Central Mediterranean route: it is estimated that while 2,6% of the migrants died in the sea during 2017, this number recently rose to 10%[251]. It is estimated that at least 13.000 people died in the Mediterranean between 2014 and 2018[252].

Those who arrive in Italy, may arrive in Sicily (63,8% of the cases), Calabria (24,4%), Campania (5,5%), Apulia (2,8%), and Sardinia (3,5%)[253]. As regards the countries of origin of the migrants, according to the statistics, following the *Arab springs,* North Africans were the most represented nationalities (in 2011, Libyans and Tunisians alone were the 84% of the total), but in 2013 and 2014 Syrians were a quarter of the total number: in the same years, indeed, Syrians were also moving to Greece. The other main countries of provenience are Somalia and Eritrea, either for the situation in their countries and for the cultural ties with Italy[254]. However, their main destinations remain Sweden, Switzerland, and Germany[255]. The other nationalities of migrants are Ivoirians, Guineas, Nigerians and Bangladeshis.

Table 3: Arrivals in Italy for year and countries of origin: data from UNHCR and Greek Ministry of the Interior, retrieved in May 2021

Year	Total number of arrivals	First country of origin		Second country of origin		Third country of origin	
2011	62.692	Tunisia	44%	Libya	38%	Egypt	2,5%
2012	13.267	Somalia	16,4%	Eritrea	10,8%	Egypt	9,2%
2013	42.925	Syria	26,3%	Eritrea	20,2%	Somalia	7,6%
2014	170,100	Syria	25%	Eritrea	20%	Mali/Nigeria/Gambia	5%
2015	153.842	Eritrea	25%	Nigeria	14,5%	Somalia	8,1%
2016	181,436	Nigeria	21%	Eritrea	11%	Guinea	7%
2017	199.369	Nigeria	9%	Guinea (Conakry)	8%	Côte d'Ivoire	8%
2018	23.400	Tunisia	22%	Eritrea	14%	Iraq	7%
2019	11.500	Tunisia	23%	Pakistan	10%	Côte d'Ivoire	10%
2020	34.154	Tunisia	33,8%	Bangladesh	11,9%	Côte d'Ivoire	7,2%
TOT	**892.685**						

[251] *Ibidem.*

[252] *Ibidem.*

[253] Castellano, V., La mappa dei porti dove sbarcano i migranti, AGI, 01 July of 2017.

[254] Riley, D., Emigh, R., Post-Colonial Journeys: Historical Roots of Immigration and Integration, Comparative Sociology, 1, 2, 2002.

[255] EASO, *Annual Report on the Situation of Asylum in the European Union,* 2020.

Table 4: Table of the arrivals in Spain, Italy and Greece per country of origin. Data from UNHCR, retrieved in May 2021

Year	Total number of arrivals	First country of origin		Second country of origin		Third country of origin	
2015	1.031.402	Syria	48,3%	Afghanistan	20,39%	Iraq	8,6%
2016	362.376	Syria	23%	Afghanistan	12%	Nigeria	10%
2017	178.500	Nigeria	10,1%	Syria	9,8%	Guinea (Conakry)	6,7%
2018	141.500	Guinea	10%	Morocco	9,6%	Mali	7,9%
2019	123.700	Afghanistan	23%	Syria	15,8%	Morocco	7,1%
2020	95.031	Tunisia	14,8%	Algeria	8%	Bangladesh	5,1%
TOT	**1.932.509**						

2.2.2 Destination Europe: the migratory routes across Asia and Africa

The origin of these great migration routes that cross the Mediterranean lies in other migration routes that, in turn, cross the Asian and the African continent. In fact migrants also come from distant countries such as Bangladesh, Afghanistan and Somalia: therefore, we will now briefly describe their paths.

The African-Mediterranean routes

As the vast majority of migrants taking the Mediterranean routes are of African origin, they follow several paths depending on their final destination: we can therefore distinguish between the *Western and Central African routes* and the *Eastern African route*[256].

Migrants coming from Senegal, Gambia, Liberia, Guinea, and Ivory Coast who decide to cross the Mediterranean may choose to take the Western Mediterranean route, thus deciding to follow the route that passes through Gao (Mali), Tamanrasset (Algeria) and then Morocco[257]. On the other hand, those who decide to try to enter Europe through the southern shores of Italy usually transit trough Niger and then they reach Libya: the border between Libya and Niger is the one in which more abuses are recorded[258]. However, the desert is also extremely dangerous, as it is the place where 40% of deaths occur during the migration process; this is due to dehydration, starvation, sickness, lack of access to medication, kidnapping and

[256] Idemudia, E., *Travelling Routes to Europe*, cit.
[257] *Ibidem*.
[258] UNHCR, *Routes towards the Western and Central Mediterranean Sea*, January 2021.

abuses by smugglers[259]. Asylum seekers from Nigeria, Mali, Burkina Faso and Benin tend to converge in Niger, although in 2016 an agreement between the EU and Niger provided for increased border controls in order to avoid the migratory pressure on Algeria and Libya[260]. According to available data, 45% of the migrants travel on their own, 39% thanks to the help of friends or family and 11% resort to the help of smugglers: this suggests that migrants may seek their help when they lack right documentation[261]. Moreover, 71% of people cross the desert via bus, 15% via car and 7% via truck[262]. As for those coming from Eastern Africa, they tend to travel across Sudan and then to Libya (sometimes even via Egypt). Sudan's capital, Khartoum, is the main migratory hub and it is where most of the smuggling networks are located; there, migrants are often abused, and sometimes are even kept in slavery conditions[263]. As noted for the Central Mediterranean route, since 2017 the situation in Libya has led the migrants to choose to depart from Egypt rather than from Tripoli[264].

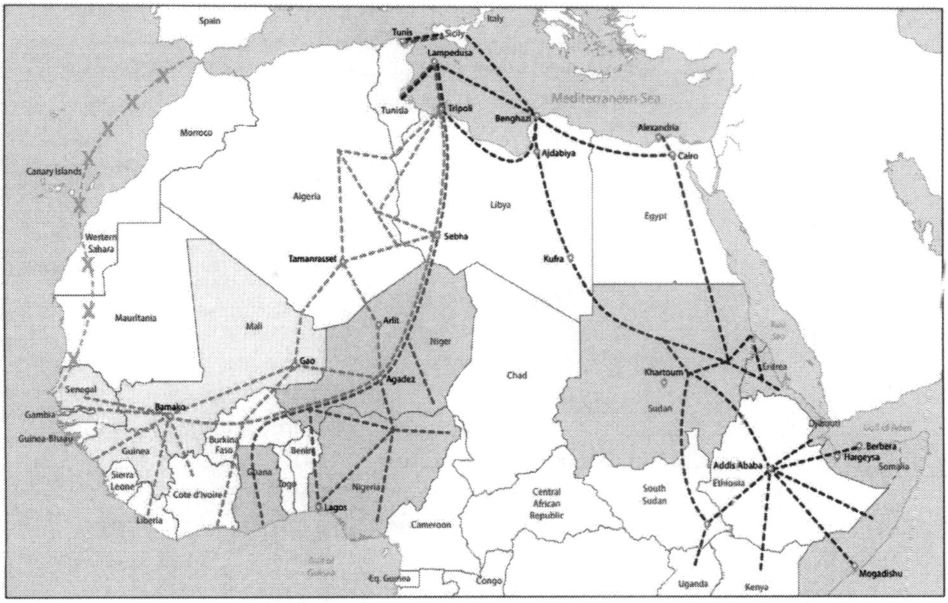

Figure 14: African routes, from, The Global initiative against transnational organized crime, Smuggled Futures: The dangerous path of the migrant from Africa to Europe, 2014.

[259] Idemudia, E., *Social experiences of Migrants,* in Psychosocial Experiences of African Migrants in Six European Countries, pp 119-135, 2020.

[260] Munsch, T., et at., *Before the desert,* Conditions and Risks on Mixed Migration Routes through West Africa, Danish Refugee Council, 2017.

[261] *Ibidem.*

[262] *Ibidem.*

[263] IOM, *Sudan 2017-9,* Migration crisis operational framework, 2019.

[264] Marchand, K., *Study on Migration Routes in the East and Horn of Africa,* Maastricht Graduate School of Governance, 2017.

The Asian routes

As seen, migrants coming as far as Pakistan, Bangladesh and Afghanistan are a large part of the total number of asylum seekers arriving – mostly – on Greek shores. Their route is as complicated as that of African migrants and in the case of both Afghans and the Pakistanis, it passes through Iran and Turkey; a different route is taken by Bangladeshis.

Migrants from Afghanistan can enter first Pakistan and then Iran, or directly Iran, depending on their country of origin or their point of departure: those entering Pakistan (via Karachi, Quetta and Peshawar) must cross the Hindukush, the desert and the steppe, while those who try to enter Iran must cross the fence which is placed at the side of the border[265]. Once in Iran, they have to reach Turkey, a passage that is controlled by organized networks of smugglers: in addition to the abuses perpetuated by local organized crime, the migrants have to face a difficult night crossing on foot of the mountain range that divides the two countries[266]. The crossing is predominantly made by groups of 50-100 people; they then cross Turkey in cars or buses[267]. A similar route is taken by Pakistanis, who usually leave from Quetta and reach Turkey in the same way; researchers have also found that a small number of Pakistanis arrive in Turkey by plane with a tourist visa[268]. Then, both Pakistanis and Afghans cross the Aegean Sea: some, as we have seen, on makeshift boats, arriving on the Greek islands; others, instead, hide in trucks and private cars, or even aboard private sailboats heading to Italy[269]. In general, during their journey, migrants travel in cars, boats, planes, buses and on foot; they usually arrive in Turkey by car or bus, cross the sea on boats and travel through Europe by walk or n rented cars, when they are not stopped by the authorities[270].

Finally, the number of Bangladeshi arrivals has increased in recent years: in fact, while in 2016 only three migrants arrived by sea were from Bangladesh, their number rose to 4645 between January and April 2017 alone[271]. Their journey usually starts in the Gulf countries, from where they fly to North-African countries: the main air routes are "Dubai-Istanbul-Tripoli, Dubai-Cairo-Khartoum-Tripoli, Dubai-Istanbul-Tunisia/Libya"[272]. There, they usually spend a year before attempting the journey to Italy; sometimes the traffickers seize their documents, forcing them to work or in

---

[265] Dimitriadi, A., *Migration from Afghanistan to third countries and Greece*, Hellenic Foundation for European and foreign policy, 2013.

[266] *Ibidem*.

[267] Kushminder, K., et al., *Irregular Migration Routes to Europe and Factors Influencing Migrants' Destination Choices,* Maastricht Graduate School of Governance, 2015.

[268] Yousef, K., *The vicious circle of irregular migration from Pakistan to Greece and back to Pakistan,* Hellenic Foundation for European and foreign policy, 2013.

[269] *Ibidem*.

[270] REACH, *Migration from Afghanistan to Europe (2014-17),* Research report, October 2017.

[271] Dockery, W., *What's behind the influx of Bangladeshi migrants to Europe?* InfoMigrants, 14th of July 2017; Al Javed, H., *Bangladesh's fatal obsession with the European dream*, Dhaka Tribune, 11th of February 2020.

[272] IOM, *Current migration trends from Bangladesh to Italy Focus on arrivals by sea,* IOM Italy Briefing, 1, June 2017.

detention centres, as happens to the other migrants[273]. In the past, Bangladeshis also travelled to Eastern European countries, from where they attempted to enter Europe; however, more Bangladeshis now leave from Libya, rather than from Turkey, Albania or Morocco[274]

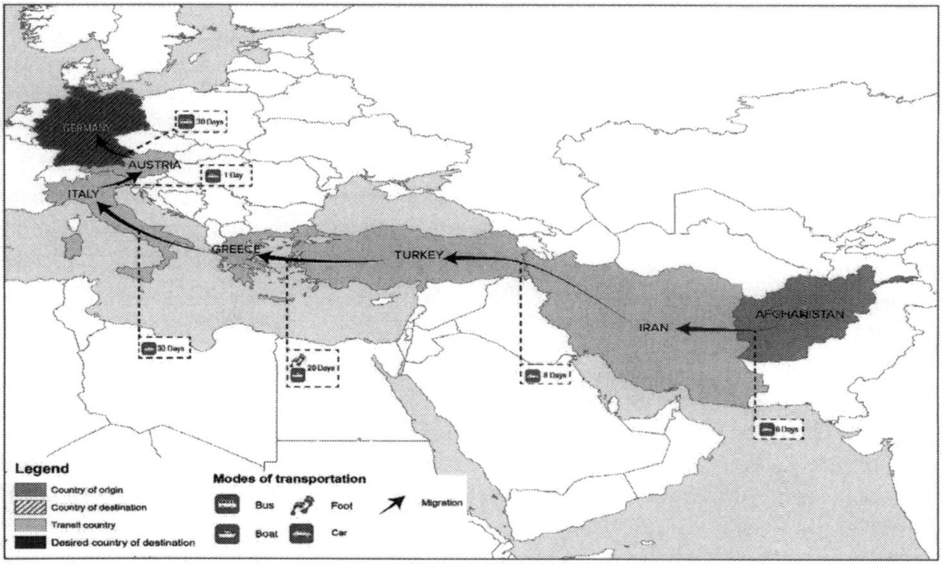

Figure 15: Route from Afghanistan to Europe (REACH, Migration from Afghanistan to Europe (2014-17), Research report, October 2017).

2.2.3 Why Europe? Migrants' main reasons to leave

When it comes to the causes of migrations, as already stated in chapter one, researchers usually distinguish between "push" and "pull" factors: push factors usually differ between countries and depend on local political and economic considerations while pull factors refer to the "reception conditions of the destination"[275]. As we shall see, both pull and push factors may vary and can differ from country to country, though they are generally similar. Regarding pull factors, for instance, a study argues that while there are mainly shared elements such as the importance of a safer life in Europe, migrants network, and democracy, there may be some elements that can vary, such as the "image" and the "reputation" of a certain country[276].

[273] Qayum, N., *Chasing the Dubai Dream in Italy: Bangladeshi Migration to Europe*, Migration Policy Institute, 5th of October 2017.

[274] Rahman, M., Kabir, M., *Moving to Europe: Bangladeshi Migration to Italy*, Institute of South Asian Studies Working Paper, 142, 2012.

[275] Kang, Y., *Refugee crisis in Europe: determinants of asylum seeking in European countries from 2008–2014*, Journal of European Integration, 43, 1, 2021.

[276] Kushminder, K., et al., *Irregular Migration Routes to Europe and Factors Influencing Migrants' Destination Choices*, cit.

Another factor could also be the migratory legislation[277]. Moreover, while economic theory suggests that the more a society is wealthy, the more it will attract foreign people especially from poorer countries, many studies argue that economic factors have a minor role in the final decision to leave[278]. In addition, according to Yoo-Duk Kang, due to constraining factors such as time pressure and exceptional circumstances, migrants may not be aware of the conditions of the places of destination[279]: interestingly, as we will see, 80% of migrants arriving from Sub-Saharian Africa to North Africa intend to stay there and do not attempt to cross the Mediterranean[280], where richer countries are located.

As far as push factors are concerned, according to several studies, violence is considered to be the most important: refugees mainly leave their countries because they are fleeing persecution and conflict[281]; we will see later how many facets violence can have. On the other hand, poverty alone is not considered a major reason to migrate, as there is not a clear correlation between this factor and the decision to migrate[282].

Another variable to be included is distance, but this has an inverse relationship with the number of migrants: the further away two countries are, the fewer migrants there will be in the destination country[283]. Finally, Kushminder et al., also stress the role of "safety along the routes, weather conditions, border surveillance and push-back policies"[284]; however, as the authors conclude that these factors lead to the change of the routes, but not to their suppression, we will not focus on them, as they do seem to prevent migrants from starting their journey. In general, migrants move for a number of reasons and not for one reason alone: we will therefore focus on the geographical area of origin to understand the local reasons of migration.

West Africa

We have previously identified violence as one of the main reasons for migration, and West Africa is no exception: Nigerian migrants in particular – but Western Africans in general – argue that growing tensions between Christians and Muslims, as well as intergenerational conflicts, land disputes and fights between families, often

[277] Idemudia, E., *Theoretical Explanations of Migrations, Mental Health, Wellbeing and Posttraumatic Stress Disorder,* in: Psychosocial Experiences of African Migrants in Six European Countries. Social Indicators Research Series, vol 81. Springer, Cham, 2021.

[278] Kang, Y., *Refugee crisis in Europe: determinants of asylum seeking in European countries from 2008-2014,* cit.

[279] *Ibidem.*

[280] Munsch, T., et at., *Before the desert, Conditions and Risks on Mixed Migration Routes through West Africa,* cit.

[281] Crawley, H, et al., *Destination Europe? Understanding the dynamics and drivers of Mediterranean migration in 2015,* Centre for Trust, Peace and Social Relations, Coventry University, Coventry, 2016.

[282] Kang, Y., *Refugee crisis in Europe: determinants of asylum seeking in European countries from 2008-2014,* cit.

[283] *Ibidem.*

[284] Kushminder, K., et al., *Irregular Migration Routes to Europe and Factors Influencing Migrants' Destination Choices,* cit.

result in murders[285]. These conflicts, however, are not localized or are not regional issues: respondents to surveys argue that groups such as the Movement of Democratic Forces of Casamance in Senegal, Boko Haram and particular fraternities such as the Nigerian Black Axe are considered a life threat[286]. The relationship between Boko Haram and migration has also been confirmed by another study, which claims that 97% of the people leaving the North-west of Nigeria move due to the activity of the terrorist group[287]. Gambia respondents also cited persecution related to the membership of a political party and unfair legal processes as reasons to move[288]. Violence was not mentioned, on the other hand, by Malians and the Nigeriens, who indicate that the lack of job prospects was the factor which motivated them to leave their country; however, in turn, the lack of employment perspective has been linked to a general situation of instability and conflict, as well as to poverty and marginalization[289]. This is also evident in the World Bank report on West Africa, in which it is stated that "social and political grievances have played a central role in motivating youth violence"[290]: therefore, violence, triggered by the poor social environment, further generates more violence, further destabilizing the region and forcing more and more people to flee. Indeed, it is important to point out that in 40% of the cases, migrants coming from this region have at least a secondary school level of education[291] thus could have greater opportunities to find a job than their compatriots.

A recent study also claims that migrants traveling the Atlantic route have referred to the desire to escape poverty and to achieve greater economic security, even though the route is widely known to be extremely dangerous, while others claim that lack of fishing was also a motivation, since this was their main activity[292]. The lack of fishing is explained by the migrants as a result of the governments' decision to sell fish resources to "foreigners", while other issues such as the lack of access to land, as well as agreements which favour European companies, are considered factors that are also having effects on the local economy[293].

Recently, the number of Congolese travelling on the Central Mediterranean route has in creased: according to the *Portail sur les données migratoires,* this is due to years of endemic violence in the country. Conflict is thus responsible of a deteriorating security situation and of severe food insecurity: little less than one million

[285] Crawley, H, et al., *Destination Europe? Understanding the dynamics and drivers of Mediterranean migration in 2015*, cit.

[286] *Ibidem.*

[287] Mukthar, S., et al., *Boko Haram and the Geopolitics of Forced Migration in Nigeria,* Journal of International Studies Vol. 14, 51-63, 2018

[288] Crawley, H., et al, *Destination Europe? Understanding the dynamics and drivers of Mediterranean migration in 2015*, cit.

[289] Munsch, T., et at., *Before the desert,* Conditions and Risks on Mixed Migration Routes through West Africa, cit.

[290] Marc, A., *The Challenge of Stability and Security in West Africa,* International Bank for Reconstruction and Development / The World Bank, 2015.

[291] *Ibidem.*

[292] Affeldt, J., Seaman, M., *A Gateway Re-opens: the growing popularity of the Atlantic route, as told by those who risk it,* cit.

[293] *Ibidem.*

refugees leaving Congo were documented from April 2020 to February 2021[294]. Most of them are hosted in neighbouring countries, especially Uganda (45%)[295], but some arrived in Greece via the Eastern Mediterranean Route (1004 in 2020)[296].

East Africa

In East Africa, two countries are among those hosting the greatest number of refugees in the world: Uganda (1.4 million)[297] and Ethiopia (805.164)[298]. In addition, the greatest number of refugees come from just one country, South Sudan (2.3 million). The other main nationalities from Eastern Africa documented by European authorities are the Eritreans and the Somalis. However, even though Somali refugees are more than 750.000 and are hosted in neighbouring countries, Somalia itself hosts refugees from Ethiopia and Yemen[299]: this gives the idea of a general instability which is also reflected in arrivals on the European shores. As it was already evident from Western Africa, every country has both different and similar patterns of migration: overall, the entire region has a low human and economic development but is also facing conflicts, political oppression and persecutions[300].

As for South Sudan, according to the studies, the main factors forcing people to flee their homes are "national conflict (62%), communal clashes (20%) and natural disasters (15%)"[301]; South Sudan, indeed, is at the centre of several disputes, especially regarding the Nile basin[302]. Indeed, as we will see in the third paragraph, in Sub-Saharan Africa, the environmental issue is becoming increasingly relevant. Nonetheless, these environmental disputes and crossed interest are also at the root of the civil war that started in 2013; ongoing peace talks are trying to come to a permanent peace[303] and to stop a war which has caused at least 383.000 deaths after only 5 years of war[304]. The widespread violence clearly affects people, but it has also caused food insecurity: indeed, it is estimated that half of the population has become food insecure due to the effects of the war[305]. Moreover, the conflict has also had an impact on the country's economy, with poverty in urban areas increasing from 45% to 70%. We will investigate the "environmental factor" further, but for now it is sufficient to say that 95% of the population relies on "climate sensitive sector such as agriculture, fisheries and forestry resourc-

[294] Portail sur les données migratoires, *Migration data in Eastern Africa,* 02 February 2021.

[295] UNHCR, *Operational portal, Refugee Situation, DRC Situation,* accessed on 04/05/2021.

[296] UNHCR, *Operational portal, Refugee Situation, Mediterranean situation, Greece,* cit.

[297] UNHCR, *Refugee Data Finder,* accessed on 04/05/2021.

[298] UNHCR, Operational portal, Refugee Situation, *Ethiopia,* accessed on 04/05/2021.

[299] Portail sur les données migratoires, *Migration data in Eastern Africa,* 02 February 2021.

[300] Marchard, K., *Study on Migration Routes in the East and Horn of Africa,* cit.

[301] Portail sur les données migratoires, *Migration data in Eastern Africa,* cit.

[302] Knopf, P., *South Sudan's Civil War and Conflict Dynamics in the Red Sea,* Special report, United States Institute of Peace, 2018.

[303] Sant'Egidio, *South Sudan: together with Sant'Egidio major step forward towards Peace,* 11th of March 2021.

[304] Specia, M., *383,000: Estimated Death Toll in South Sudan's War,* The New York Times, 26th of September 2018.

[305] Putoto, G., *Environmental Vulnerability: South Sudan's Endgame,* ISPI online, 23rd of March 2018.

es" and that South Sudan is listed among the "worst performing countries in the Climate change vulnerability index 2017"[306].

The Sudanese region of Darfur faces the same levels of violence of South Sudan, with attacks, harassment by authorities, and violence of militias being the main reason for migrants to leave[307]. Clearly, the level of violence prevents people from finding a job, and even when they do, Darfuri may suffer discrimination: the situation is particularly hard for younger generations, which describe themselves as "hopeless" and regard the migration to Europe as their only option and their unique chance to escape this "systemic persecution"[308]. The rest of the country, on the other hand, is facing inter-tribal violence, especially in the Abyei territory, as well as environmental problems such as soil degradation, drought, and desertification[309]. Climate change is thus exacerbating conflicts over lands and resources: 80% of the rural population, indeed, relies on the agriculture – and thus, on climate conditions – for their living[310].

Other conditions that pushes migrants to move – especially the youngest ones – is the lack of perspectives, as have already seen with the West African countries: forty percent of the population is below 15 years of age but, according to IOM, the government has not enacted policies designed to cope with the needs of the youths[311]. The lack of perspective is also linked to the economic situation, which continues to deteriorate due to the years of conflict and civil war[312].

Somalia is also an example of a complex mix of both environmental issues and violence: IOM estimates that in 2020 alone, 1.6 million people were affected by the effects of adverse climatic events such as severe drought and famine[313]. To the 30th of November 2021, 665.104 refugees had left Somalia[314] and 3 million Somalis were Internally Displaces Persons[315]; Somalia, however, also faced the consequences of the lack of a central government until 2012 but continues to struggle with the destabilizing chronic insecurity, which is also linked to poverty[316]. Indeed, according to a 2018 report, Al-Shabaab was the "deadliest group of the Continent": the terrorist movement targets government institutions, civilians, IOs, and even the international missions such as AMISOM[317]. Moreover, clan disputes over lands and politi-

---

306 *Ibidem.*

307 Buchanan-Smith, M., *Darfuri migration from Sudan to Europe: from displacement to despair*, Humanitarian Policy group, August 2018.

308 *Ibidem.*

309 IOM, Migration Crisis Operational Framework, *Sudan*, cit.

310 *Ibidem.*

311 *Ibidem.*

312 *Ibidem.*

313 IOM, *Climate Change Drives Displacement of Thousands of Vulnerable People in Somalia*, 4th of January 2021.

314 UNHCR, Operational portal, Refugee Situation, *Horn of Africa Somalia situation*, accessed on the 10th of January 2022.

315 UNHCR, Operations, East and Horn of Africa and Great Lakes, *Somalia*, accessed on the 10th of January 2022.

316 Marchard, K., *Study on Migration Routes in the East and Horn of Africa*, cit.

317 Mbiyozo, A., *Fleeing terror, fighting terror: the truth about refugees and violent extremism*, Institute for Security Studies, 2018.

cal power, further aggravate the situation[318]; indeed, Somalis also move to escape persecutions related to clan and political affiliations[319]. The lack of access to food, medical services, and healthcare, provoked by the continuous violent situation, provides another stimulus[320].

Eritrea has also faced famine and drought in recent years, but the main driver of migration among youths Eritreans is considered to be the mandatory 18 month-military service, which, however, ends up lasting much longer; this, together with political oppression and poor economic conditions, forces people to leave the country[321]. Nonetheless, it is important to note that among those who leave the country, 27% have a university level education, 37% a secondary level and little less than 5% a teacher training[322]. However, when fleeing Eritrea, they risk being shot at the border or risk consequences for their families[323]. The recent tensions in the Tigray region in Ethiopia are further aggravating the situation of the Eritrean refugees hosted in the area[324].

Finally, as the other countries in the Horn of Africa, Ethiopia faced the effects of the 2011 drought that affected the regions, as already mentioned; this provided a stimulus for Ethiopians to leave their country, as have the ethnic divisions and other socio-economic factors[325]. In 2013 the Ethiopian Government imposed a ban on migration, but this does not prevent many Ethiopians to irregularly leave the country[326].

Middle East and Asia

The threat of violence is a common driver of migration from Iraq, Syria and Afghanistan: just to introduce the issue, it is sufficient to say that 79% of the Iraqi migrants would have left their country even if offered a job and that among the returnees only 20% would not try to reach Europe again[327].

The origin of the crisis that is forcing both Iraqis and Syrians to leave their homeland lies in the *Arab springs* that led to the civil war in Syria and to the insurgence of the Islamic state in both Iraq and in Syria: while protests were dividing Syria in 2011, indeed, the US were leaving Iraq and by 2013 ISIS had begun to be a real

---

[318] *Ibidem.*

[319] UNHCR, IOM, *A long and winding road,* Background Paper, Regional Conference on Refugee Protection and International Migration: Mixed Movements and Irregular Migration from the East and Horn of Africa and Great Lakes Region to Southern Africa, Dar Es Salaam, 2010.

[320] Mbiyozo, A., *Fleeing terror, fighting terror: the truth about refugees and violent extremism,* cit.

[321] Marchard, K., *Study on Migration Routes in the East and Horn of Africa*, cit.

[322] Kibreab, G., *The national service/Warsai-Yikealo Development Campaign and forced migration in post-independence Eritrea,* Journal of Eastern African Studies, 7,4, 630-649, 2013.

[323] *Ibidem.*

[324] UN News, *Deep concern for thousands of Eritrean refugees 'scattered' in Ethiopia's Tigray,* 26th of March 2021.

[325] Adugna, G. *Migration patterns and emigrants' transnational activities: comparative findings from two migrant origin areas in Ethiopia*, CMS 7, 5, 2019; Marchard, K., *Study on Migration Routes in the East and Horn of Africa*, cit.

[326] ILO, *Addressing the Root causes of Migration in Ethiopia,* accessed on the 4th of May 2021.

[327] IOM, *Iraq, Migration drivers and reasons for migrating to Europe,* 2018.

challenge to the Iraqi state[328]. ISIS was declared defeated in 2019, but between the 2013 and 2019 the terrorists had ruled over a wide territory, causing the displacement of millions of people, both in Syria and in Iraq; however, while Iraq is now hosting a huge number of Syrian refugees, and less Iraqis are migrating, Syrians continue to leave their country[329]. UNHCR, indeed, documented 5.687.007 Syrian refugees so far[330].

State and non-state violence, therefore, seems to be the main driver of the Middle East migration, as said before: according to the studies, indeed, 28% of the migrants interviewed in Greece reported that they had to leave due to the terrorist activity of ISIS, though not only in Syria and Iraq but also in Afghanistan and Yemen[331], while state persecution is also cited by Syrians and Pakistanis[332]. According to the work of Crawley et al., indeed, 87% of the Syrians, 88,6% of the Afghans, and 93% of the Iraqis arrived in Greece claim that they moved due to "conflict, human rights abuses and persecution"; in addition, even those who were not targeted argued that the level of violence had become intolerable and a threat for their families[333].

As for the Afghanis, they mentioned their Hazara ethnic identity as a threat, because they experienced discrimination and lack of rights even when they fled to neighbouring Iran[334]. Moreover, according to the REACH report, war and conflict (20%), lack of job opportunities (5%) and tribal conflict (less than 5%) were the primary reasons to leave Afghanistan, while lack of job opportunities were the secondary reasons to migrate for the 20% of the interviewed migrants[335]. Importantly, this report was published before the Taleban's return to power.

Pakistan too is among the ten world countries per numbers of emigrants: so far, 6.3 million people left the country, a sixth of which only in 2015[336]: among the pull factors that lead Pakistanis to Europe, studies reported availability of educational scholarships (30%), migrants' networks (21%) and job opportunities (20%)[337]. As for push factors, 39% of the respondents declared that lack of job opportunities was the main reason, 12% cited financial problems, 14% lack of quality education and 12% reunification with friends or family[338]. It is also interesting to say that among migrants who want to reach EU, 64% obtained at least a bachelor and 42% would stay in Pakistan if given better educational opportunity, a job (58%) and a safer situation (46%)[339].

---

[328] Chambers, Richard L., et al., *"Iraq"*. Encyclopedia Britannica, 11th of January 2021.

[329] UNHCR, *Regional Strategic overview 2020-2021, Regional Refugee & Resilience Plan (3RP)*, 2019.

[330] UNHCR, Operational portal, Refugee Situation, *Syria regional Refugee Response*, accessed on the 10th of January 2022.

[331] Crawley, H. et al, *Understanding the dynamics of migration to Greece and the EU: drivers, decisions and destinations,* MEDMIG Research Brief No.2, 2016.

[332] *Ibidem.*

[333] *Ibidem.*

[334] *Ibidem.*

[335] REACH, *Migration from Afghanistan to Europe (2014-17),* cit.

[336] IOM, *Pakistan: survey on drivers of migration,* 2020.

[337] *Ibidem.*

[338] *Ibidem.*

[339] *Ibidem.*

Similar considerations are made by Bangladeshi migrants: Bangladesh, indeed, is the sixth country per number of migrants worldwide: statistics reveal that 7.5 million Bangladeshis live abroad. The country is experiencing slow economic growth, overpopulation, and natural disasters[340]: thus, key drivers of migration are the necessity to find a better job and livelihoods (73%) but also to increase the social status (11%)[341]. Moreover, the three main causes behind Bangladeshis' migration are unemployment (33%), family reunification (27%), and access to a community abroad (25%)[342]. Nonetheless, 91% of the migrants would anyway remain in their homeland if they were given a job[343]. Moreover, 28% of the Bangladeshi migrants claim that they choose a destination for its availability of jobs and better salaries[344].

The impact of Covid-19

The impact of Covid-19 could have long-lasting effects on the societies from which migrants come; surveys in West Africa, for instance, report that the pandemic is acting as a threat-multiplier, i.e. it is reducing the access to work, increasing worries and stress, and reducing the availability of goods[345]. According to the surveys, indeed, 60% of the interviewed refugees could not meet their basic needs; 40% were selling their goods to buy food and 11% had to move to less safe accommodations[346]. In addition to the "push factors" represented by the economic fallout, an interesting "pull factor" motivating young Africans to leave their country is the perceived shortage of jobs in Europe, influenced by the news of thousands of deaths in the European Continent[347]. Moreover, as the majority of the migrants is facing difficulties when moving across borders, an increased need for smugglers has been reported: furthermore, smugglers are also choosing more dangerous routes[348]. It seems therefore that the more the pandemic will impact economy, the more migrants will risk their lives on dangerous routes. Indeed, while in May 2020 it seemed that the number of arrivals on the European shores was decreasing[349], in January 2021 it was clear that numbers had increased, compared to 2019 (at least in Italy and Spain)[350]. In conclusion, it is no surprise that, according to data, many new arrivals came from countries affected by economic difficulties rather than by conflicts[351]: again, this clearly demonstrated that migration is likely to increase due to the economic difficulties triggered by the pandemic, considering also the global reduction of remittances[352].

---

[340] Yousef, K., The vicious circle of irregular migration from Pakistan to Greece and back to Pakistan, cit.

[341] IOM, *IOM Releases Results of Largest Survey on Drivers of Migration in Bangladesh,* 14th of July 2020.

[342] IOM, *Bangladesh, survey on drivers of migration and migrants' profile,* 2020.

[343] *Ibidem.*

[344] *Ibidem.*

[345] Litzkow, J., *West Africa: How the Pandemic Reshapes Migration*, ISPI online, 17th of September 2020.

[346] UNHCR, *Routes towards the Western and Central Mediterranean Sea,* cit.

[347] Affeldt, J., Seaman, M., *A Gateway Re-opens: the growing popularity of the Atlantic route, as told by those who risk it,* cit.

[348] *Ibidem.*

[349] IOM, *Mixed migration flows in the Mediterranean,* 2020.

[350] European Commission, *Migration statistics update: the impact of COVID-19,* 29th of January 2021.

[351] *Ibidem.*

[352] *Ibidem.*

2.3 "Climigrants" and "ecodemics"?

2.3.1 Climate changes' effects

According to the Intergovernmental Panel on Climate Change (IPCC), "scientific evidence for warming of the climate system is unequivocal"; moreover, NASA claims that it is extremely likely (more than 95% of probability) that the current global warming is the "result of human activity since the mid-20th century"[353]. However, even though the IPCC was founded in 1988, the issue of climate change has only recently entered the political and legislative debate[354].

The beginning of modern efforts to mitigate the effects of human activity on climate must be set in 1952, when London was covered by a huge smog cloud and the English legislator consequently had to restrict the burning of domestic coal[355]. During the 1950s, American mothers became concerned about the effects of the nuclear irradiation on their babies' health, but it was not until 1970 that, after the Santa Barbara oil spill and the Cuyahoga River burning, the US Environmental Protection Agency was founded[356]. It was becoming evident that industries impacted not only on their local regions, but also on the global environment: for instance, invasive species, carried by long-distance transport, were beginning to threaten local species, with direct effects on human health[357].

For centuries, societies were influenced by climatic events: they had poorer individual life outcomes, but had little impact of the environment; by contrast, in the era of the *Anthropocene,* humans have improved individual outcomes but are causing systematic changes to the global environment. However, the "double-edged sword of science and industry"[358] has given humans the illusion to control Nature, but the effects of this perceived control are now having an impact on human society: the very symbol of the globalized society, air transport, "has made every local outbreak a new disease"[359], spreading worldwide new viruses resulting from the crossing of animal-human border. We will see later how spill-over infections are becoming more frequent due to the invasions of animal habitats by humans and how migration can also be linked to the effects of human-induced climate change.

In NASA's opinion, the evidence of climate change can be found in the rise of temperatures connected to the oceans' warming, the shrinking of the ice sheets, the glacial retreat, the decreased snow cover, the sea level rise, the declining arctic sea ice, the increasingly common sea events and the ocean acidification[360]. The rise in

---

[353] NASA, *Climate Change: How Do We Know?* Accessed on the 6th of May 2021.

[354] Ebbs, S., *Experts call for climate change debate to move beyond settled science and focus on action,* ABC News, 22nd October 2020.

[355] Brooke, J., *Growth beyond Limits: 1945 to Present,* in Climate Change and the Course of Global History: A Rough Journey, Studies in Environment and History, 529-558, Cambridge University Press, Cambridge, 2014.

[356] *Ibidem.*

[357] *Ibidem.*

[358] *Ibidem.*

[359] *Ibidem.*

[360] NASA, *Climate Change: How Do We Know?* Cit.

temperatures is caused by the constant emission of greenhouse gases (water vapor, nitrous oxide, methane, and carbon dioxide) produced by human activities, which trap the solar radiation within the atmosphere, thus warming the planet[361]. In addition to the proved effects of greenhouses gases emissions, studies also demonstrated that carbon dioxide concentration has never been higher for millennia (see figure 4), thus further confirming that human activities are responsible for the current global warming[362]. Even though we cannot give a precise percentage, there is wide agreement among scientists on the anthropogenic origin of global warming[363]. Therefore, we will mainly focus on the effects of climate change on European society.

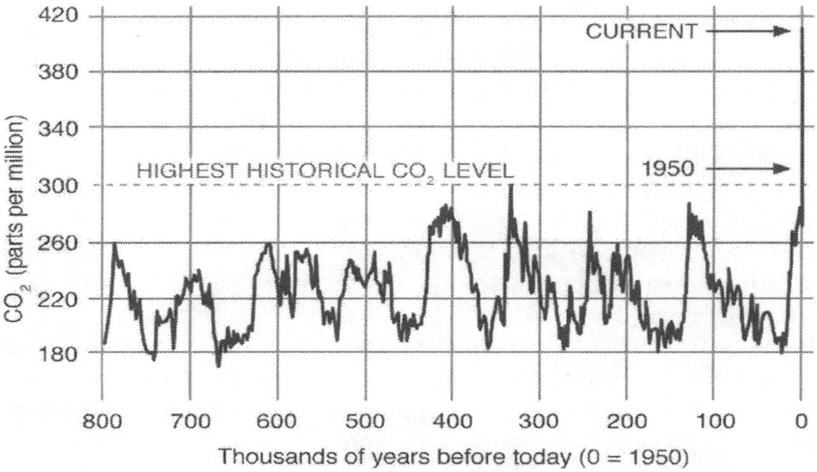

Figure 16: Carbon dioxide levels throughout history, NASA, Carbon dioxide, accessed on 11/05/2021.

Effects on the seas

First of all, as global temperatures are increasing, Arctic and Antarctic ice sheets and glaciers are melting and sea levels are rising[364], with effects on coastlines such as flooding and erosion[365]. The sea level rise is expected to have consequences on all low-lying coasts in absence of adaptation efforts[366]: however, this is not only true for tropical islands, but also for several kilometres of European coasts. Recently, for instance, the German Federal Court of Justice ruled that the 2019 Climate

---

[361] NASA, *The causes of Climate Change*, accessed on 06/05/2021.

[362] NASA, *Carbon dioxide*, accessed on the 11th of May 2021.

[363] NASA, *Scientific consensus: Earth climate is warming*, accessed on 06/05/2021; Herring, D., *Isn't there a lot of disagreement among climate scientists about global warming?* NOAA, Climate.gov, 3rd of February 2020; Powell, J., *Scientists Reach 100% Consensus on Anthropogenic Global Warming*, Bulletin of Science, Technology & Society, 37, 4,183-184, 2017.

[364] Nerem, R. et al., *Climate-change-driven accelerated sea-level rise detected in the altimeter era*, PNAS February 27, 115, 9, 2022-2025, 2018.

[365] European Commission, *Climate change consequences*, accessed on the 6th of May 2021.

[366] *Ibidem*.

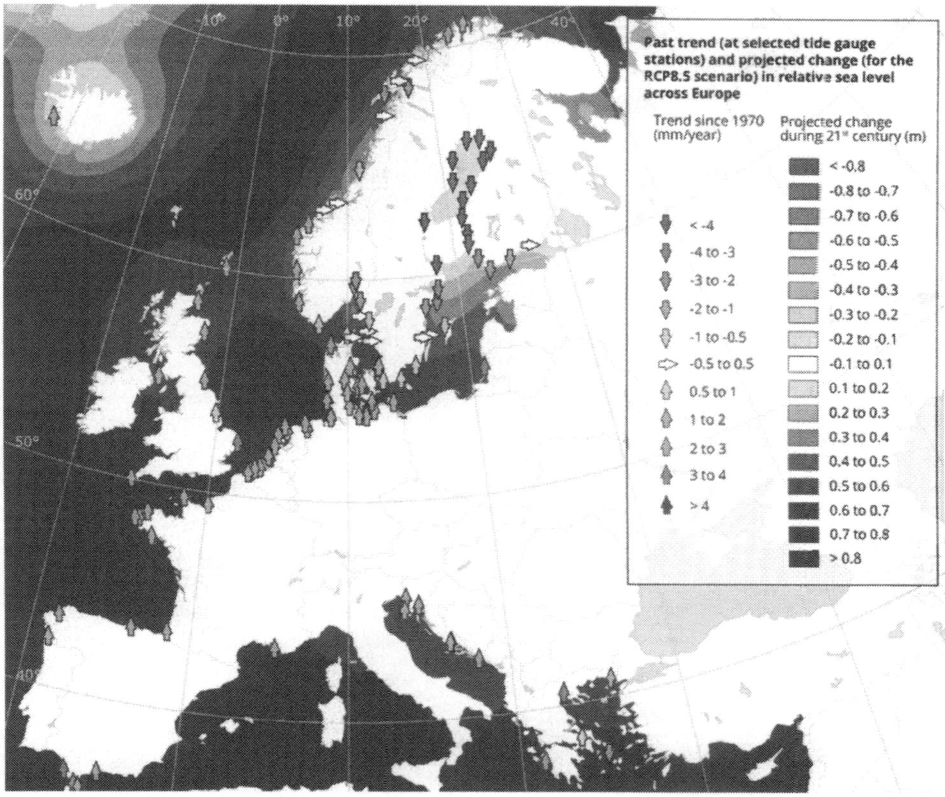

Figure 17: Projected change in the European costs. From: European Environment Agency.

Change Law does not make sufficient provisions for cutting carbon emissions, as stated by a young woman living on a small island in the north of the country: in her opinion, indeed, the effects of climate change will soon hit the island, as also reported by scientists. According to studies, indeed, sea level could rise up to 80 cm by the end of the century in both Southern England and Northern Germany[367]. The largest share of flood-related damage, however, is expected in Germany, France, Italy, the Netherlands, Belgium and the UK[368]. This will have an impact on the economy and communities living in these areas, which will be forced to migrate[369]. Moreover, the number of people exposed to coastal floods annually is expected to increase by fifteen and thirty times by 2100 and three European cities, Naples, Marseilles, and Athens, are among the twenty cities that will be most affected by flood damage by 2050[370].

[367] Grinsted, A., et al, *Sea level rise projections for northern Europe under RCP8.5*, Clim Res, Vol. 64, 15-23, 2015.

[368] *Ibidem.*

[369] Bosello, F., et al., *Economic impacts of climate change in Europe: sea-level rise*, Climatic Change, 112, 63-81, 2012.

[370] *Ibidem.*

Physical impacts	European regions*					
	Southern Europe	Central Europe South	Central Europe North	British Isles	Northern Europe	EU
Physical impacts as estimated by the agriculture model						
Yield change (%)†						
2.5 °C	0	5	−1	−9	37	3
3.9 °C	−12	5	−3	−11	39	−2
4.1 °C	−4	3	2	15	36	3
5.4 °C	−27	−3	−8	19	52	−10
Physical impacts as estimated by the river flooding model						
People affected (1,000s/y)‡						
2.5 °C	46	117	103	12	−2	276
3.9 °C	49	101	110	48	9	318
4.1 °C	9	84	119	43	−4	251
5.4 °C	−4	125	198	79	−3	396

Figure 18: Ciscar, J., Physical and economic consequences of climate change in Europe, PNAS, 108, 7, 2678-2683, 2011.

	Scenario			
	2.5 °C	3.9 °C	4.1 °C	5.4 °C
World population in 2100 (10^{12})	10.4	15.1	10.4	15.1
World GDP in 2100 (10^{12}, 1990 US$)	235	243	235	243
CO_2 concentration (ppm)	561	709	561	709
Δ Temperature (°C)*				
World	2.4	3.1	2.3	3.1
EU†	2.5	3.9	4.1	5.4
Southern Europe	2.6	4.1	4.3	5.6
Central Europe South	2.4	3.9	4.4	6.0
Central Europe North	2.3	3.7	4.0	5.5
British Isles	1.6	2.5	3.2	3.9
Northern Europe	2.9	4.1	3.6	4.7
Δ Precipitation (%)*				
EU†	1	−2	2	−6
Southern Europe	−7	−15	−13	−28
Central Europe South	2	−2	−4	−16
Central Europe North	3	1	6	−1
British Isles	−5	−2	10	5
Northern Europe	10	10	19	24
Sea level rise (high climate sensitivity) (cm)	49	56	51	59

Figure 19: Projected temperatures depending on the world mean temperatures. From: Ciscar, J., Physical and economic consequences of climate change in Europe, PNAS, 108, 7, 2678-2683, 20115.

The effects of global warming on the seas are not limited to rising sea levels; in recent years, the oceans have absorbed between one-fifth and one-third of carbon dioxide emissions, causing the acidification of the seas[371]. As for Europe, this is true for both the Mediterranean Sea[372] and the Atlantic Ocean[373]. The acidification of the sea has an impact on the ecosystems of the seas and on biodiversity[374] but it has also effects on human health: it causes respiratory problems because of the aerosolization of toxins; it impacts on human well-being as it causes the loss of habitats; it reduces the possibility of obtaining medicines and natural products that are important for human health; above all, it reduces the quantity and nutritional composition of seafood, thus leading to malnutrition, especially in coastal communities[375].

Extreme events

Between 1980 and 2019, extreme events, such as "temperature extremes, heavy precipitation and droughts" caused the death of 92,119 people in Europe and eco-

[371] Sabine, C., et al., *The Oceanic Sink for Anthropogenic CO2,* Science, 305, 5682, 367-371, 2004.

[372] Hassoun, A., et al., *Acidification of the Mediterranean Sea from anthropogenic carbon penetration, Deep Sea Research Part I: Oceanographic Research Papers,* 102, 1-15, 2015.

[373] Rios, A., et al., *Decadal acidification in the water masses of the Atlantic Ocean,* PNAS August 11, 112, 32, 9950-9955, 2015.

[374] Guinotte, J., Fabry, V., *Ocean Acidification and Its Potential Effects on Marine Ecosystems, New York Academy of Sciences,* 1134, 320–342, 2008.

[375] Falkenberg, L. et al., *Ocean acidification and human health,*Int J Environ Res Public Health, 17, 12, 4563, 2020.

nomic losses for 11.1 billion per year; worryingly, according to the IPCC, these extreme events will become even more frequent in the future [376]. Moreover, extreme droughts are projected to increase in severity, both in the moderate emission scenarios and in the extreme emission ones; Southern and Western Europe will be the regions with the heavier impact, while Northern Europe will nonetheless experience more drought events in both spring and summer[377]. However, Northern Europe is the region that would experience the highest temperature increase if temperatures in Europe will rise by 2.5°C on average by 2100; on the other hand, if the average global temperature increase were 5.5°C, Southern and Central Europe would be the region with the largest increase, 6°C[378]. In this case, Northern Europe will experience more intense rainfall (+24%) and Southern Europe will experience less precipitation (-28%, see also Figure 5)[379].

These events will affect several human activities, above all agriculture, which is dependent on temperature and water. According to Ciscar et al., in the best case scenario, Southern and Central Europe would experience low yield losses, while in the worst case scenario the region would see an important reduction in production; Northern Europe, on the other hand, will be favoured by rising temperatures in any case[380]. The Final report of the European Cooperation in Science and Technology states that the Pannonian area would also be the most affected in terms of agriculture[381].

On the other hand, heavier rainfall is expected all over Europe; this does not mean that it will rain more, but that there will be greater intensity of precipitation and longer dry periods in Europe[382]. Moreover, a recent study has also stated that there will be an intensification of short-term heavy rainfall exceeding high thresholds, leading to "flash floods"[383]. River floods, indeed, are estimated to increase and affect between 250,000 and 400,000 people per year by 2080[384].

Europe's first "climigrants"?

According to statistics, 47% of Europeans say that climate change represents a major threat to society and 41% of young Europeans think that they will have to move to another country because of global warming[385]. This is particularly evident among Southern and Eastern Europeans, 90% of whom already feel the impact of cli-

---

[376] European Environment Agency, *Economic losses from climate-related extremes in Europe,* 16th of March 2021.

[377] Spinoni, J., et al., *Will drought events become more frequent and severe in Europe?* Int. J. Climatol., 38: 1718–1736, 2018.

[378] Ciscar, J., *Physical and economic consequences of climate change in Europe*, PNAS, 108, 7, 2678-2683, 2011.

[379] *Ibidem.*

[380] *Ibidem.*

[381] COST Action 734, *Impact of climate change and variability on European agriculture'*, edited by Simone Orlandini, Pavol Nejedlik, Firenze University Press, Firenze 2012.

[382] European Environment Agency, *Heavy precipitation in Europe,* 23rd February 2021.

[383] *Ibidem.*

[384] Ciscar, J., *Physical and economic consequences of climate change in Europe,* cit.

[385] European Investment Bank, *2019-2020 EIB climate survey,* accessed on the 6th of May 2021.

mate change on their lives[386]. Northern Europeans, on the other hand, are among those who do not perceive the impact of climate change, despite recent temperature records (Norway, for instance, recently experienced its warmest November since 1900)[387]. Scientists agree that the effects of current climate change are already particularly evident in the Arctic and in northern regions, where "surface air temperatures are increasing at more than twice the rate of the global average"[388]. The effects are already visible in the lands of northern Europe and are already changing the habits of an indigenous population, the Sami.

The Sami are a nomadic people who have been living in Arctic Scandinavia and Lapland for centuries: although there is no formal census based on ethnicity[389], it is widely accepted that there are around 100,000 Sami in this part of Europe[390]. The historical occupation of the Sami has been reindeer herding, which makes them highly dependent on the environment and drives them to lead a nomadic life. Nowadays, however, only 10% of the Sami are engaged in this occupation: this is due to sociological reasons and climate change. Indeed, more and more Sami are migrating in search of new opportunities and this is also due to the effects of climate change: we can thus consider the Sami as the first European "climigrants", migrants leaving their homes due to the effects of global warming.

The current effects of climate change on this nomadic population are twofold. From a strict point of view, since the Sami are reindeer herders, we must consider the said impact of rising temperatures on the environment, "food, predators and habitats"[391] as well as "changing vegetation"[392]. A warmer climate is increasing the risk of wildfires, as well as stress among humans and animals, animal-borne diseases, and mental health problems related to the fear of future[393]. On the other hand, climate change is also making these northern lands more habitable, and an increasing number of people are deciding to settle there, also thanks to the Norwegian policy of relocating Syrian refugees in this area[394], aimed at populating the region: a *"folkevandring"* (notably, the same word used by Germans to describe the *Migration period*) that is being studied by many scholars[395] and is changing the demography of the region.

---

[386] *Ibidem*.

[387] Davis, S., *Europe's autumn 'the hottest on record amid world's warmest November'*, 07th December 2020.

[388] Yamanouchi, T., Takata K., Rapid change of the Arctic climate system and its global influences - Overview of GRENE Arctic climate change research project (2011–2016), Polar Science, 2020.

[389] Kelman, I., Naess, M. W., *Climate change and displacement for Indigenous Communities in Arctic Scandinavia*, Center for International Climate and Environmental Research, Oslo, 2013.

[390] Kelman, I., Naess, M. W., *Climate Change and Migration for Scandinavian Saami: A Review of Possible Impacts,* MDPI, 2019; Furber, M., Evengard B., Nilsson, M., *Facing the limit of resilience: perceptions of climate change among reindeer herding Sami in Sweden*, Global Health Action, 2019.

[391] Kelman, I., Naess, M. W., *Climate Change and Migration for Scandinavian Saami: A Review of Possible Impacts,* cit.

[392] *Ibidem*.

[393] Jakkola, J., *The Holistic Effects of Climate Change on the Culture, Well-Being, and Health of the Saami, the Only Indigenous People in the European Union,* Curr Environ Health Rep.,5,4, 401-417, 2018.

[394] *Ibidem*.

[395] Naguib, N., Humanitarian Pluralism: *The Arctic Passage in an Age of Refugees*, International Journal of Middle East Studies, Cambridge University Press, 2016.

Refugees, however, are not the only newcomers: while younger Sami generations are already leaving Lapland in search of "education, larger settlements, or more varied livelihoods"[396], it is "resource extraction [which] will bring major changes to Arctic Scandinavia"[397]. Indeed, the Sami now face competition from these new industries, which are reducing lands for reindeer herding[398]. However, despite the negative effects produced by the warm climate, the Sami are not fiercely opposing the exploitation of their land: they just want to be consulted so that all stakeholders can benefit from the extraction of resources. In this sense, migration associated with climate change "is not inevitable"[399] even if it is linked to successful mediation.

Nevertheless, previous social changes and assimilation policies have contributed to the loss of the Sami's heritage, making them even more vulnerable to the negative effects of climate change. Moreover, although the Covid-19 pandemic did not affect cross-border herding, it raised concerns about the price of reindeer meat and the activities of the Sami in general[400].

In conclusion, so far, the migration of Sami communities is not due to the primary effects of climate change, but rather to the "social and governance factors that inhibit adaptation [to climate change] and which in turn force people to migrate"[401]. It is to be expected, though, that the lingering effects of climate change will continue to affect the region, allowing new resource extraction activities and forcing the Sami to leave Lapland.

2.3.2 Facing the effects of climate change from outside

The migratory movement of the Sami is not the only migratory movement triggered by climate change: Europe is already witnessing the arrival of migrants leaving due to environmental reasons. Rinsing sea levels, coastal flooding, salinization of low-lying agricultural lands, increased intensity of tropical cyclones and storms, changes in rainfall regimes with consequences for agricultural production, higher temperatures and more frequent fires, changes in chemistry affecting agricultural production and maritime ecosystems, are "heavily felt in poorer countries" and, in the words of the European Parliament, are "likely to influence migration drivers"[402]. In some African and Asian countries out and in-migration are already indirectly triggered by climate changes through wages and agricultural productivity: therefore,

---

396 Kelman, I., Naess, M. W., *Climate Change and Displacement for Indigenous Communities in Arctic Scandinavia*, cit.

397 *Ibidem.*

398 Kelman, I., Naess, M. W., *Climate Change and Migration for Scandinavian Saami: A Review of Possible Impacts*, cit.

399 *Ibidem.*

400 The Arctic Council, *The impact of Covid-19 on Saami communities*, 2020.

401 Kelman, I., Naess, M. W., *Climate Change and Displacement for Indigenous Communities in Arctic Scandinavia*, cit.

402 Kraler, A., et al., *Climate Change and Migration: Legal and policy challenges and responses to environmentally induced migration*, Policy Department for Citizens' Rights and Constitutional Affairs Directorate-General for Internal Policies, European Parliament, 2020.

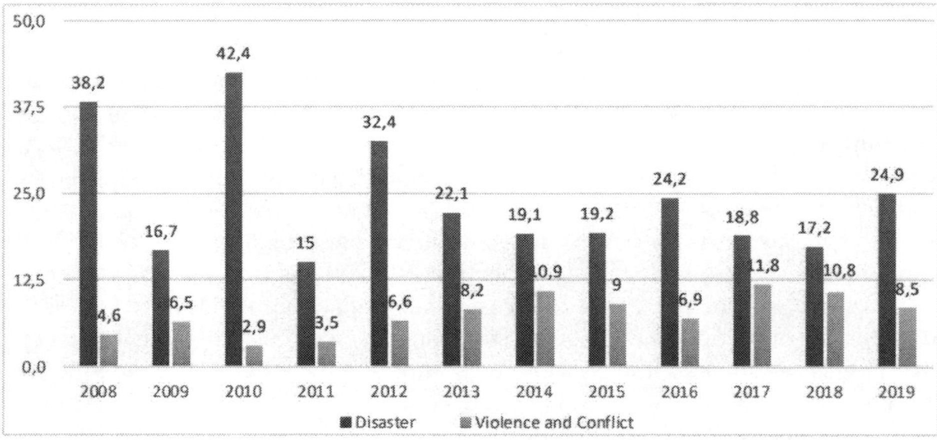

Figure 20: Disaster-and conflict-related internal displacement, from Kraler, A., et al., Climate Change and Migration: Legal and policy challenges and responses to environmentally induced migration, Policy Department for Citizens' Rights and Constitutional Affairs Directorate-General for Internal Policies, European Parliament, 2020.

even though "individuals may still view their migration decisions primarily as economically motivated, deteriorating environmental conditions can be linked to migration dynamics by analysing the impact of environmental change on the economy"[403]. While it is clear that extreme natural events can have an impact on the local community with direct effects on crop production, there are also indirect links between climate change and conflict-related displacement. We will therefore focus on them.

According to Black et al., we can list five drivers of migration: economic, political, demographic, social and environmental factors[404]. However, the presence of drivers of migration does not mean that migration will actually take place, as it is the result of an individual's decision; age, gender, level of education, wealth…these factors play a role in the decision, as so do family characteristics, barriers and facilitators. Nonetheless, climate change could influence both economic, political, demographic, and social drivers, even though there is no agreement on this issue (at least, not for all factors). We will therefore focus again on the countries of origin of migrants arriving in Europe, looking for evidence of the effects of climate change on migration decisions.

West Africa

Africa – along with Asia – is considered to be the Continent that will be most affected by climate change. Moreover, it is also important to consider that 65% of the Sub-Saharan workforce is linked to rain-based agriculture[405] and that the num-

[403] *Ibidem*.

[404] Black, R., et al., The effect of environmental change on human migration, Global Environmental Change, 21,1, S3–S11, 2011.

[405] Freeman,L., *Environmental Change, Migration, and Conflict in Africa: A Critical Examination of the Interconnections*, Journal of Environment & Development 26,4, 2017.

ber of undernourished people has increased by 45% since 2012: Sub-Saharan Africa is one of the regions that have been severely affected by drought in recent years[406]. However, as mentioned above, we have to consider that the decision to migrate comes from a number of reasons, but research is revealing how environmental factors are playing a role in migration from several West African countries: the drying up of Lake Chad, for instance, with its consequences, could be considered as a driver of outmigration[407]. In Mali, extreme climatic conditions contributed to the decline of Lake Faguibine, which was completely dry for 21 years between 1976 and 2008; this led 200,000 people to abandon their livelihoods, transforming the traditional economy[408]. Moreover, river flows in Niger, Senegal, and Gambia have decreased by 25 to 60% over the last 30 years. In addition, increasing outbreaks of malaria have been linked to unusually heavy rains, after which they usually occur[409]. Importantly, consequences of the same unusually heavy rains are also the floods that have been documented in the region in recent years, leading to the displacement of hundreds of thousands of people and to major crop losses, thus increasing the prices of rice, corn, and wheat[410]. The 2007 flood in Ghana, for instance, caused the loss of 260,000 metric tons of cereals and food items, while 330,000 people were forced to leave their homes[411].

Furthermore, the effects of climate change have also been noticed in the Gulf of Guinea, where 56% of the coastlines of Benin, Cote d'Ivoire, Senegal and Togo are already being eroded, with consequences for crops growing near the coasts[412], but also for the migration of fish stocks[413]; extreme fishing by foreign trawlers, as said before, is thus worsening a situation aggravated by climate change. We must bear in mind that 4.8 million people in West Africa rely on fishing for their livelihoods[414]. Extreme weather events, therefore, could be behind the decision to leave, as they have an impact on agriculture, affecting the subsistence economy but also increasing prices. Moreover, new studies are linking the effects of climate change to the other main driver of outmigration in the region: conflict and violence.

According to several authors there may be a connection between climate change and violence in the crucial region of Lake Chad, which borders Cameroon, Chad, Niger and Nigeria. Since 2009, clashes between armed groups have destabilized the area and created 2.5 million refugees: Janani Vivekananda, senior adviser for Climate

[406] United Nations Climate Change, *Climate Change Is an Increasing Threat to Africa,* 27th of October 2020.

[407] Freeman,L., Environmental Change, Migration, and Conflict in Africa: A Critical Examination of the Interconnections, cit.

[408] United Nations Environment Programme, *Livelihood security: climate change, migration and conflict in the Sahel,* 2011.

[409] United Nations Climate Change, *Climate Change Is an Increasing Threat to Africa,* cit.

[410] United Nations Environment Programme, *Livelihood security: climate change, migration and conflict in the Sahel,* cit.

[411] *Ibidem.*

[412] *Ibidem.*

[413] Lovei, M., *Climate Impacts on African Fisheries: The Imperative to Understand and Act,* World Bank Blog, 11th of November 2017.

[414] Muggah, R., *In West Africa, Climate Change Equals Conflict,* Foreign Policy, 21st of February 2021.

Change and Peacebuilding at Adelphi, argues that this is due to increasing and unpredictable rainfall which, with its effects on agriculture, is increasing tensions and exacerbating problems, triggering competition for increasingly scarce natural resources[415]. Young people are thus forced to leave their lands, but as they lack the education and skills necessary to find a job in urban areas, they are pushed into poverty: poor and marginalized, they easily become prey to radical groups such as Boko Haram[416].

Moreover, even before the emergence of Boko Haram, a terrorist group that is terrorizing the region, there was growing competition for land between opposite groups, such as herders and farmers. Herders, who are almost 25 million in Sahel, are dependent on the weather, which is changing: dry seasons are longer, and rainfalls less regular, thus posing the problem of reduced access to water[417]. Competition is also increasing on the coast, both among fishermen[418] and between fishermen and coastal guards[419].

Therefore, it is becoming evident how climate change, by exacerbating existing fragilities, increases the attractiveness of financial incentives provided by armed and terrorist group such as Boko Haram[420], but also triggers small conflicts that could even lead to escalation[421].

East Africa

When discussing the main drivers of migration from Eastern Africa we cited the case of South Sudan, where 95% of the population is dependent on agriculture, fisheries and forestry resources, and is therefore highly susceptible to climate change. We also looked at Somalia, where drought has displaced millions of people. Indeed, Eastern Africa is no new to droughts: in 2011, 13.5 million people in Somalia, Kenya and Djibouti were suffering the effects of droughts of past years[422] and in 2019, when another drought occurred, the number rose to 15.3 million[423]. Moreover, other droughts occurred between 2017 and 2015[424] and previously in 2009, 2008, 2006, 2005, 2001, 2000 and in the 1990s[425]. It must be said that drought is endemic in the Horn

[415] Vivekananda, J., *Climate Change and the Conflict Trap in Lake Chad,* ISPI online, 19th of March 2020.

[416] Onya, C., *Climate Change and Conflict in Nigeria: The Boko Haram Challenge,* American International Journal of Social Science, 4, 2, 2015.

[417] Muggah, R., *In West Africa, Climate Change Equals Conflict,* cit.

[418] United Nations Environment Programme, *Livelihood security: climate change, migration and conflict in the Sahel,* cit.

[419] Muggah, R., *In West Africa, Climate Change Equals Conflict,* cit.

[420] *Ibidem.*

[421] United Nations Environment Programme, *Livelihood security: climate change, migration and conflict in the Sahel,* cit.

[422] International Federation of Red Cross and Red Crescent Societies, *Drought in the Horn of Africa. Preventing the next disaster,* 2011.

[423] Goldenberg, M., *Drought in the Horn of Africa is threatening 15 Million People,* UN Dispatch, 5th of August 2019.

[424] Internal Displacement Monitoring Center, *Recommendations for addressing drought displacement in Ethiopia,* 2021.

[425] International Federation of Red Cross and Red Crescent Societies, *Drought in the Horn of Africa. Preventing the next disaster,* cit.

of Africa and that local societies have adapted and migrated across the region for centuries, maintaining emergency wells and remembering where water and pasture were; however, it must also be said that several non-climate factors are changing the region, affecting the livelihoods of its inhabitants[426]. Borders prevent migration, while enclosures restrict access to wells, and conflicts over land are exacerbating the situation, as clashes between herders often turn into serious conflicts[427]. The tensions already present in the Horn of Africa are thus exacerbated by the worsening effects of climate change. According to the IOM there were 1.7 million internally displaced persons (IDPs) in Ethiopia at the end of 2019: 489,359 of them were due to climate-related events, while the rest had to leave their homes because of conflicts[428].

As for Somalia, 3 million people are actually displaced inside the country, and UNHCR argues that they had to leave their homes "mainly due to conflict, but increasingly due to climate-related shocks such as severe droughts and flooding": 70% of the 1.3 million new Somali refugees in 2020 were driven by flooding[429]. In addition, in the past thirty years, the country has not only experienced severe droughts, but also fifteen tropical storms and cyclones as well as flash floods: experts claim that these are part of the stronger storms caused by climate change and warmer ocean temperatures[430]. When cyclone Gati arrived in Somalia in November 2020, 46,000 people had to leave their homes[431].

Moving north, statistics show that there are 1.67 million IDPs in South Sudan: 15% of them due to natural disasters, 20% due to communal clashes, while 62% had to leave because of the national conflict[432].

South Sudan was also important to mention because of recent research indicating that conflict is more likely to take place where floods and droughts are more probable[433]. Once again, we find a nexus between climate change and conflict: among other things, research from 2014 shows that droughts and livestock losses make people more susceptible to recruitment by Al-Shabaab, a Somali terrorist group[434]. Moreover, as in the case of West Africa, changing pastoral mobility patterns, linked to changing environmental conditions, could degenerate into conflicts among herders; in addition, low-scale conflicts, also related to climatic shifts, may be exacerbated by politics, as in the Darfur conflict (2003-2009), when tensions between farmers and nomads were used to co-opt the nomads in the Janjaweed militia, or in Kenya, where the Moi regime used land grievances between pastoralists and farmers for political purposes[435]. In general, Assistant Professor van Balen and

[426] *Ibidem.*

[427] *Ibidem.*

[428] Portail sur les données migratoires, *Migration data in Eastern Africa,* cit.

[429] UNHCR, *UNHCR airlift supplies to assist thousands of Somalis displaced by Cyclone Gati,* 18th December 2020.

[430] *Ibidem.*

[431] *Ibidem.*

[432] Portail sur les données migratoires, *Migration data in Eastern Africa,* cit.

[433] Putoto, G., *Environmental Vulnerability: South Sudan's Endgame*, cit.

[434] Abshir, S., *Climate change and security in the Horn of Africa: can Europe help to reduce the risk?* European Institute of Peace, 2020.

[435] *Ibidem.*

Senior Research Office Mobjörk claimed that there is a link between hot temperatures and violent conflicts in East Africa; for instance, "in Somalia, abnormally high temperatures and drought cause herders to sell more of their livestock than under normal conditions, causing an oversupply of low-quality animals that depresses prices in local consumer markets. This increases the risk of violent conflicts between groups by triggering economic price shocks"[436]. Livelihood losses, therefore, contribute to conflict as that they exacerbate tensions over resource sharing and push people to join armed groups. However, this does not mean that violence "automatically follows when peoples' livelihoods are under stress"; nonetheless, slow environmental change would give people more time to negotiate access to resource and develop cooperation[437]. On the other hand, a rapid climate change is increasing tensions and damaging the economy, forcing people to migrate before developing adaptation strategies.

Middle East and Asia

An analysis by Zhang et al. confirmed that the Middle East region has been experiencing a warming trend since the 1970s: cold days are decreasing, while warm days are intensifying, with a sudden sharp increase in the 90s (recorded in South and Central Asia too)[438]. The climate warming is also reflected in the number of droughts, the most severe of which, according to Barlow et al., were in 1999-2001 and 2007-2008 (see also Figure 7)[439]. Moreover, in the Middle East region, climate is epected to become drier in the future and the Fertile Crescent may even disappear by the end of the century: this is worrying as the region is already experiencing significant water stress[440].

Drought, on the other hand, is linked to food insecurity. The UN reports that in the Middle East rain-fed crops are highly influenced by droughts; current climate change is thus decreasing yields and vegetation in pasture lands, which in turn affects cattle[441]. Moreover, this also has an impact on households and the overall economy, as for instance, agriculture accounted for about 25% of Syria's GDP before the 2007-2010 drought hit the country. However, agriculture production declined throughout the Middle East after the 2010 drought[442].

In the period between 1998 and 2006, Afghanistan experienced the most severe drought in its climate history; this caused crops and cattle losses, and also in-

---

[436] van Baalen, S., Mobjörk, M., *A coming anarchy? Pathways from climate change to violent conflict in East Africa*, Stockholm International Peace Research Institute & The Swedish Institute of International Affairs, 2016.

[437] *Ibidem.*

[438] Zhang, X., et al., *Trends in Middle East climate extreme indices from 1950 to 2003,* Journal of Geophysical Research, 110, 22, 2005.

[439] Barlow, M., et al., *A Review of Drought in the Middle East and Southwest Asia*, Journal of Climate, 29(23), 8547-8574, 2016.

[440] *Ibidem.*

[441] Haamed, M., *Drought and food security in the middle east: An analytical framework,* Agricultural and Forest Meteorology, 281, 2020.

[442] *Ibidem.*

Kyrgyzstan	2009
Turkmenistan	
Uzbekistan	2000/01
Tajikistan	2000/01, 2008
Afghanistan	1969, 1971–73, 2000–02, 2006, 2008, 2011
Pakistan	1999–2003
Syria	1999–2000, 2008–10
Lebanon	
Israel	1999
Saudi Arabia	
Iraq	1969–71, 1998–2001
Iran	1964, 1999–2001
Jordan	1999, 2000
Kuwait	
Yemen	1969–71, 1975, 1977

Figure 21: List of the most severe droughts in the Middle East region, Barlow, M., et al., A Review of Drought in the Middle East and Southwest Asia, Journal of Climate, 29(23), 8547-8574, 2016.

creased prices. In 2018, 43% of the rural population (9.8 million people) were in a food crisis; one year later, 287,000 people became IDPs because of the drought and an estimated 13.5 million people were food insecure; in addition, floods caused by unusual rainfalls damaged villages and crop. This is a serious problem as 85% of the population relies on agriculture[443]. Recently, the UN argued that 70% of the population has been forced to leave their homes because of climate change[444].

Droughts in the region are expected to be more frequent in the future, due to the effects of climate change, and this is likely to exacerbate socio-political tensions[445]. There is a large body of literature which links the outbreak of the Syrian Civil War, and the subsequent destabilizing situation in the Fertile Crescent, with the 2007-2010 drought mentioned before[446]: we will now focus on this issue.

As early as 2009, a study argued that "shifting rainfall patterns, spreading desertification, and falling agricultural productivity are likely to undermine rural livelihoods, worsen job prospects in rural areas and accelerate migration to urban areas. This could put a strain on services in cities and lead to increased resentment of existing refugee populations". The authors of the above-mentioned study then focused their attention on Syria, where they stated that "the 2007/8 drought caused significant hardship in rural areas [...] In the northeast of the country, a reported 160 villages have been entirely abandoned and the inhabitants have had to move to urban areas"[447]. Two years later, in 2011, civil war broke out. Later, other authors, as mentioned, tried to find a link between the Syrian Civil War and the drought: indeed, the entire region, in the words of the authors of the afore-mentioned study, Brown and Crawford, has

[443] Prívara, A., Prívarová, M., Nexus between Climate Change, Displacement and Conflict: Afghanistan Case, Sustainability, 11, 5586, 2019.

[444] Prívara, A., Prívarová, M., Nexus between Climate Change, Displacement and Conflict: Afghanistan Case, Sustainability, 11, 5586, 2019.

[445] Cook, B., et al., *Spatiotemporal drought variability in the Mediterranean over the last 900 years*, J. Geophys Res Atmos, 121, 5, 2060-2074, 2016.

[446] Gleick, P., *Water, Drought, Climate Change, and Conflict in Syria,* Weather, Climate, and Society, 6, 3, 331-340, 2014; Feitelson, E., Tubi, A., *A main driver or an intermediate variable? Climate change, water and security in the Middle East*, Global Environmental Change, 44, 39-48, 2017; Brown, O., Crawford, A., *Rising Temperatures, Rising Tensions - Climate change and the risk of violent conflict in the Middle East,* International Institute for Sustainable Development, 2009; Saha, S., How climate change could exacerbate conflict in the Middle East, Atlantic Council, 14th of May 2019; Fischetti, M.. *Climate Change Hastened Syria's Civil War,* Scientific American, 2nd of March 2015; Kelley, C. et al., *Climate change in the Fertile Crescent and implications of the recent Syrian drought,* PNAS, 112, 11, 3241-3246, 2015.

[447] *Ibidem*.

already experienced decades of bloody wars and climate change could only exacerbate the situation in several ways[448]. Prior to the Syrian Civil War, in fact, the Levant was already struggling with "scarce water, food insecurity and erratic economic growth, each of which could be exacerbated by climate change" as it could increase competition for scarce water resources, intensify food insecurity, worsen poverty and social instability, force migration and trigger further tensions, increase militarization and lead to further distrust of the West, in the words of Brown and Crowford[449]. Again, it is interesting to note that this escalation has indeed taken place in Syria. However, before going any further, we should also include in the discussion neighbouring Iraq, whose western part is highly dependent on the Euphrates river; most importantly, its upstream course is used by both Syria and Turkey for irrigation purposes[450]. The two Gulf War influenced water management and, as Iraq was already experiencing social tensions since 2003, the 2007-2010 drought was intertwined with an already destabilized socio-economic environment[451]. During the drought, agricultural production dropped (wheat production fell by 47% and barleycorn production by 67%) and several herders lost more than 80% of the livelistock; finally, prices increased and so did malnutrition and diseases among children. Even though the Syrian government appealed to the UN, only a small proportion of the funds were received, and soon 1.5 million people left the north-east for the cities, while an equivalent number of Iraqi refugees did the same, triggering an increase in the urban population: the illegal tent camps where they settled had limited access to infrastructure and refugees received little or no help from the Government and could not find work. These cities became the centres of the social tensions that developed and sowed the seeds of the Syrian civil war; meanwhile, in neighbouring Iraq, the same drought forced people to abandon their homes, increasing resentment against the Government and creating the basis for the spill-over of the conflict to Northern Iraq. Lastly, it is interesting to quote the response of a woman, when asked about a possible correlation between climate change and revolution: "Of course. Drought and unemployment were important in pushing people towards revolution. When the drought happened, we managed to handle it for two years, and then we said, 'It's enough'[452].

Remarkably, other studies also find possible links between changing climatic conditions and the *Arab springs*. It has to be said that rising food prices were only an "aggravating factor" driving people to demand change and that the riots that took place in the MENA region in 2011 should not be described as "bread riots". However, the protests in Algeria were indeed triggered by higher prices for basic goods, people were fed up with inflation in Tunisia and even in Egypt households spent 40% of their income on food due to rising inflation[453]. Climate change with its

---

[448] *Ibidem*.

[449] *Ibidem*.

[450] Feitelson, E., Tubi, A., *A main driver or an intermediate variable? Climate change, water and security in the Middle East*, cit.

[451] *Ibidem*.

[452] ; Kelley, C. et al., *Climate change in the Fertile Crescent and implications of the recent Syrian drought*, cit.

[453] Johnstone, S., Mazo, J., *Global Warming and the Arab Spring*, in The Arab Spring and Climate Change, A Climate and Security Correlations Series, Edited by Caitlin E. Werrell and Francesco Femia,

consequences, therefore, acted as a "threat multiplier" also in North Africa[454]. In the region, climate change has contributed to inflation, triggering social discontent, already caused by water scarcity; both desertification and increasingly unpredictable and unusual rainfalls have contributed to and accelerated migration flows, which on the other hand have also exacerbated conflicts and damaged the capacity of states to meet the needs of the population[455].

As for Afghanistan, by "enhancing the attractiveness of poppy production, deteriorating climatic conditions are exacerbating insecurity and instability in Afghanistan"; the cultivation of opium, indeed, is associated with conflict and violence, as it is highly profitable and finances anti-governments groups, thus contributing to State fragility[456]. Moreover, degradation and huge demographic growth are triggering competition for land, fuelling historical conflicts such as that between sedentary Hazara and Pashtun herders. Moreover, according to the researcher Price, conflicts over natural resources are also part of a broader geopolitical framework[457]; therefore we cannot argue that climate change is fuelling violence in Afghanistan, but we can observe that, as in other regions of the Middle East and North Africa, it is exacerbating pre-existent tensions[458].

Pakistan, which borders Afghanistan, is among the areas most affected by climate change due to the effects of drought and floods; in particular, as in many of the countries analysed, 45% of the labour force is employed in agriculture[459]. Climate change, migration and violence in Pakistan have been defined as having a "causal relationship": surveys have showed that droughts, heat waves, floods, and disease are affecting people's livelihoods. Moreover, overpopulation, land scarcity, economic competition, and migration have a clear link to changing climatic conditions, and it is clear from the surveys that socioeconomic challenges, dwindling resources, migrants and water shortages are the most common causes of conflicts in the region[460].

Lastly, as for Bangladesh, we must say that climatic factors are only one of the main drivers of migration, though it has been said that the country will be one of the most affected by climate change due to its geographical characteristics. As early as 2003, land and water scarcity were the main drivers of migration; the same study stated that increasing rural-to-urban migration, triggered by floods and droughts,

---

Center for American Progress, Stimson, The center for Climate and Security, 2013.

454 *Ibidem.*

455 Werz, M., Hoffman, M., *Climate Change, Migration, and Conflict*, in The Arab Spring and Climate Change, A Climate and Security Correlations Series, Edited by Caitlin E. Werrell and Francesco Femia, Center for American Progress, Stimson, The center for Climate and Security, 2013.

456 Prívara, A., Prívarová, M., *Nexus between Climate Change, Displacement and Conflict: Afghanistan Case*, cit.

457 Price, R., *Climate change as a driver of conflict in Afghanistan and other Fragile and Conflict Affected States,* K4D Helpdesk Report 527, Institute of Development Studies, Brighton, UK 2019.

458 *Ibidem.*

459 Fahad, S., Wang, S., *Climate change, vulnerability, and its impacts in rural Pakistan: a review,* Environmental Science and Pollution Research, 27, 1334-1338, 2020.

460 Ali, F., et al., *Climate Change-Induced Conficts in Pakistan: From National to Individual Level,* Earth Systems and Environment, 2, 573-599, 2018.

would drive between 3 and 10 million people out of Bangladesh by 2040[461]. In the period between 2005 and 2011 alone, however, more than 3.5 million Bangladeshis left their country[462]. In 2007, 9 million people were made homeless by floods[463] and in 2020, the worst flood since 1988 inundated nearly one million homes[464], affecting 2.4 million people [465] and inundating a third of Bangladesh's territory[466].

In conclusion, as studies agree that worsening climate conditions will negatively affect Bangladesh[467], we may argue that migrants arriving in Europe can be regarded as the first migrants who left also because of changing climatic conditions.

2.3.3 Are we dealing with "ecodemics"?

In the first paragraph, we observed that the twenty-first century pandemics were or are *zoonoses,* that is to say infections or diseases transmitted by animals to humans: bubonic plague, SARS, MERS, HIV, Ebola, flu, and Covid-19 are all examples of zoonotic diseases[468]. These types of infections are caused by contact with wild species, such as bats, monkeys, camels, and pangolins, to name just a few; however, even though humans hunted animals for millennia, according to WWF, consumption of bushmeat is increasing dramatically and so is wildlife trafficking[469]. Both practices risk transmitting viruses to humans both because of the consumption of bushmeat (Ebola's outbreaks have been linked to the hunting and slaughter of infected meats) and because wild animals are often herded close together in unhygienic conditions nearby human dwellings. Moreover, the fact that exotic animals are frequently sold together in markets increases the possibility of genetic recombination[470]. Furthermore, it is also food insecurity – triggered by the effects of anthropogenic climate change too – that often drives people to hunt exotic animals such as chimpanzees, hippos, and antelopes, especially in Africa[471]. So-called "wet markets", one of which is also considered to be the starting point of the current Covid-19 pandemic, are known for selling wild animals such as scorpions, crocodiles, monkeys, bats, rodents, insects and so on; however, these markets also feed millions of poor people in both Africa and Asia[472]. On the other hand, we must also say that

[461] Mahmoeei, B., Parris, B., *Climate change and internal migration patterns in Bangladesh: an agent-based model,* Environment and Development Economics, 17,6, 763-780, 2012.

[462] Bundeszentral fur politische Bildung, *International Migration from Bangladesh,* 30th of November 2015.

[463] BBC, *Rivers and flooding: Case study: Bangladesh,* accessed on 10/05/2021.

[464] NASA, *Intense flooding in Bangladesh,* accessed on 10th of May 2021.

[465] OCHA, *Daily Noon Briefing Highlights: Bangladesh*, 27th of July 2020.

[466] Dhaka Tribune, *One-third of Bangladesh under water in worst floods in a decade*, 15th of July 2020.

[467] Mahmoeei, B., Parris, B., *Climate change and internal migration patterns in Bangladesh: an agent-based model,* cit.

[468] Jeffries, B., *The loss of Nature and the rise of pandemics*, WWF, 2020.

[469] *Ibidem.*

[470] *Ibidem.*

[471] *Ibidem.*

[472] Vidal, J., *Destroyed Habitat Creates the Perfect Conditions for Coronavirus to Emerge*, COVID-19 may be just the beginning of mass pandemics, Scientific American, 18th of March 2020.

wildlife goods are a source of profit for many guerrilla groups such as Al Shabab and Boko Haram, among others[473]; interestingly, we have previously linked the emergence of these groups to changing climate conditions.

It is important to note that spill-over is also facilitated by the disruption of ecosystems, i.e. the disruption of forests, intense mining, and increasing urbanization: together with rapid population growth, these actions are bringing humans in closer contact with animals as never before[474]. However, urbanization is also increasing population density, facilitating the spread of a possible new virus. As pathogens do not respect species boundaries, humans are facilitating the outbreak of new viruses in the sense that they are creating new artificial habitats while destroying natural ones; in shantytowns on the edge of cities, people live in close contacts with species that have adapted to the new environment, such as rodents, bats and birds, which are the most likely to transmit new diseases to humans[475].

Importantly, land-use change, deforestation, and modification of natural habitats are also considered responsible for the outbreak of almost half of zoonoses[476]; however, deforestation and disruption of natural habitats not only indirectly harm humans by creating the possibility of new virus spill-over, but also contribute to climate change. In fact, as scientists have shown[477], deforestation prevents trees from absorbing carbon dioxide produced by human activities; moreover, not only does deforestation impact the environment by eliminating a potential aid against the CO2emissions, but it also produces carbon dioxide, when cut trees are burnt. Indeed, between just two years, 2015 and 2017, the loss of tropical forests contributed to the creation of 4.8 billion tonnes of CO2 per year (8-10% of global emissions)[478]: for instance, if forests burning were a state, it would rank third among countries with the highest level of emissions[479]. Furthermore, deforestation also alters rainfall patterns, further drying and warming existing forests[480].

Conclusion

We began this chapter by asking whether it was possible that anthropogenic climate change could be the factor behind both the pandemics of the twenty-first century and the migrations we read about every day in the newspapers; however, in the course of the analysis, we focused in depth on both the effects of the current Covid-19 pandemic and the migratory routes and other factors that drive people from their homes. The decision to adopt such a holistic perspective was motivated

[473] Brashares, J., et al., *Wildlife decline and social conflict. Policies aimed at reducing wildlife-related conflict must address the underlying causes*, Science Magazine, 345, 6159, 376-378, 2014.

[474] Vidal, J., *Destroyed Habitat Creates the Perfect Conditions for Coronavirus to Emerge, COVID-19 may be just the beginning of mass pandemics*, cit.

[475] *Ibidem*.

[476] Jeffries, B., *The loss of Nature and the rise of pandemics*, cit.

[477] Shukla, J., et al, *Amazon Deforestation and Climate Change*, Science, 247, 4948, 1322-1325, 1990.

[478] Dean, A., *Deforestation and climate change*, Climate Council, 21st of August 2019.

[479] Gibbs, D., et al., *By the Numbers: The Value of Tropical Forests in the Climate Change Equation*, World Resource Institute, 4th October 2018.

[480] Nunez, C., *Deforestation explained*, National Geographic, 7th of February 2019.

by two purposes. First of all, it was necessary to analyse our historical era in all its facets in order to compare *our* period with the *Migration* period, which will be the topic of the next chapter. Moreover, since this work also aims to provide possible policies to address the issue of climate change, an approach was needed to analyse the effects of global warming on everyone's daily life. Pandemics, economic hardships, increasing waves of migration, conflicts: these events could become the norm, forcing people to leave their homes, or prompting states to adopt restrictive measures to face new viruses.

In our analysis, however, we found no evidence that climate change has caused the current Covid-19 pandemic, nor that it is driving people to European shores. Nonetheless, we have seen that the effects of climate change, especially droughts and floods, are already causing the displacement of people to many rural countries in Africa and South-east Asia, but not yet in Europe. We have also acknowledged that habitat disruption and deforestation, one of the drivers of anthropogenic climate change, are also prerequisites for a zoonotic disease to evolve into a pandemic. On the other hand, we can still argue that the Covid-19 pandemic could have broken out anyway, but we have also observed that the likelihood of a zoonotic spill-over is increased by the disruption of habitats and this seems to be confirmed by the numerous epidemics/pandemics we have witnessed over the last twenty years.

In addition, we considered that both pandemics, drought and flood-induced loss of fields and crops lead to food insecurity and job losses; we also understood how these factors, in turn, could be "push factors", that is to say they can be incentives for people to leave their country. Moreover, we learnt that food insecurity itself can trigger disputes and conflicts, and even lead people to join armed groups; we also recognized that conflicts could trigger migration, but migration can cause conflicts too, such as between shepherds and farmers in Africa and Asia. So, migration and conflicts can cause each other.

Finally, we have seen how displaced people seek refuge in towns and villages, where they risk being marginalized and then co-opted by armed groups such as Boko Haram or Al Shabaab, thus fuelling violence and triggering further displacement of people. Alternatively, they also face difficulties in finding a job, as happened to the many Syrian refugees who later demonstrated and revolted against the Assad regime. It is important to note that IDPs and refugees, when they converge in towns and cities, mainly settle in the suburbs and in makeshift shelters; by doing so, they contribute to the progressive urbanization of the environment. Moreover, by living in poor conditions, they are at risk of contracting zoonoses, not least because, as we have seen, wild meat is also an important part of the diet of millions of people in South-East Asia and Africa. In addition, progressive urbanization not only contributes to the loss of the environment (thus causing climate change), but also to the rapid spread of zoonotic epidemics, as people live concentrated in small areas and in close contact with animals.

In conclusion, we can argue that the links we draw between these possible events do not mean that climate change *actually* cause conflict, migration, and pandemics. However, in this study we have seen that due to climate change, trends already visible could become the norm in the near future. Even in Europe, the effects of climate change could cause huge social changes, as is evident from this analysis;

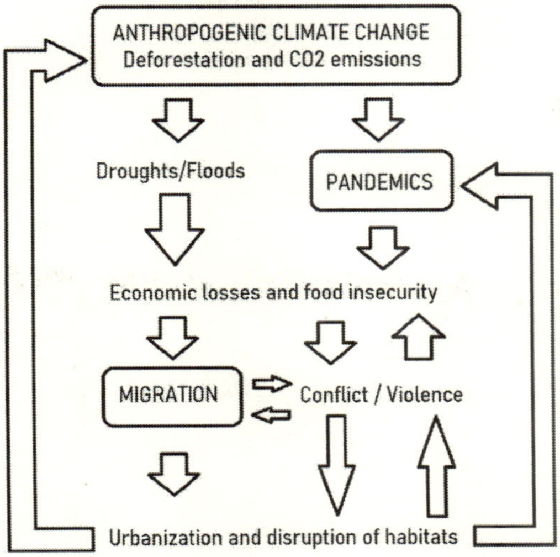

Figure 22: Hypothesized links between Climate Change, Pandemics and Migrations

moreover, Europe could also face new waves of migration, triggered by worsening and more frequent droughts, floods and conflicts. We must also consider that the vast majority of asylum seekers and refugees are not in Europe, but rather in Africa, South America and the Middle East; however, as climatic conditions will worsen especially in Africa and the Middle East, migrants may converge in Europe. In addition, as urbanization proceeds, zoonotic pandemics may become more frequent, triggering further destabilization.

Global historians, therefore, need to grasp historical trends, in order to understand what the future holds: this will be the goal of the next chapter. Indeed, we believe that comparing the *Migration period* and our own era can help predict possible developments and draw the necessary conclusions to provide possible policies that can help mitigate the effects of climate change.

III. Are we living in another Migration Period?

The current research stems from a question: are we living in another period of migration? Can we compare the early Middle Ages with the contemporary era? Can we find similarities between the events of that historical period and our own time? Do people move for the same reasons and could the Justinian plague have caused the same effects as the Covid-19 pandemic? Ultimately, can we learn from the past and provide policies that can meet the momentous challenges we will face in the near future?

3.1 Past, Present, Future

3.1.1 Two pandemics, one common trend?

In the first chapter we defined the period of the Justinian plague as a 'caesura' between the Roman period and the Middle Ages: this definition was motivated by the fact that the pandemic left its mark on the society of the time, not only because of its high death toll, but also because of its socio-economic effects. Today, although we are still dealing with the Covid-19 pandemic, we can already say that it will change contemporary society; for example, in such a short time we have learned to work and study from home, and we have also had to face social estrangement, abruptly changing our habits. However, we could argue that even though the first outbreak of Covid-19 was recorded less than two years ago, we have already vaccinated 60% of the world's population. Such a historical comparison, then, is not only relevant to capture similarities and common trends in humanity's response to pandemics; it is also useful for predicting possible evolutions. The task is made more urgent by the issue of anthropogenic climate change.

As we have seen, Covid-19 was not the first pandemic of the twentieth century, but it will certainly be remembered for its impact on our society because we were unable to learn from past epidemics/pandemics; however, in recent years, we have learned that spill-over of zoonotic diseases could become more frequent in the future if we continue to destroy natural habitats. Sars-CoV-3 may already be hiding in some tropical forest, waiting for a bulldozer to chop down a bat's tree. At the same time, we have also seen that changing climatic conditions interact with migrations which, in turn, we linked with the spread of diseases.

It is important to note that plague is a zoonotic disease, as is Covid-19, and is transmitted to humans in the same way as the latter. Moreover, although it is transmitted between humans in different ways, it probably spread due to migration flows or through trade routes, which at the time linked people from different parts of the world. Despite the differences, therefore, the Justinian plague and the Covid-19 pandemic had similar impacts on the societies of their times. In this sense, comparing two distant pandemics can help us understand how a society can avoid heavy impacts at both the macro and micro levels: the comparative work of global historians thus becomes extremely important in such a globalised society where a small microorganism can make states with millions of inhabitants decide to impose a lockdown.

Our analysis will therefore begin by comparing the different phases of the two pandemics.

We do not know how the Justinian plague arrived in Pelusium, although, as the city was a major port, we can assume that it travelled by ship, the "principal means of long-distance transport at the time[1]. In contrast, with regard to Covid-19 and the other pandemics of the 21st century, there is little doubt that aircraft contributed to the spread of the viruses, hence the "travel ban" imposed by several countries on flights from China in the early 2020s. However, as a disease usually needs an incubation period before symptoms appear, the virus can spread undetected until it is first detected; hence travel restrictions serve to prevent new outbreaks rather than to prevent the arrival of the virus. To give just one example, 759,493 people from China entered the United States in the period between December 2019 (when the virus likely emerged) and February 2020 (when the travel ban was issued)[2], thus likely unknowingly bringing the virus with them. The plane, in fact, "made every local outbreak a new disease"[3], as mentioned earlier; the ship, on the other hand, can even be considered a form of quarantine because of the duration of the trip, even if it did not prevent the spread of the disease. So while the virus may have taken less than a year to reach Constantinople, the capital of the Byzantine Empire, it took just under a day to get from Wuhan to New York. This gives the idea of a globalised world in which events are 'accelerated' and so are trends, hence once again the importance of comparison to predict possible developments.

Once the plague arrived in Constantinople in 542, it spread throughout the Mediterranean and beyond; the virus probably travelled mainly along trade routes, the routes of globalisation at that time. This statement is reinforced by the fact that several chronicles of the time claim that some regions were untouched by the pandemic, while they claim that cities along rivers or by the sea experienced several outbreaks. Similarly, we can imagine that economic connections may have played a role in the outbreaks in the early part of the Covid-19 pandemic: in fact, a study dating back to 23 January 2020[4] – the day the travel ban was issued for the Chinese city of Wuhan – claimed that the risks of importing cases from China based on the Europe-Wuhan networks were higher for the UK (39%), France (24%), Germany (15%), Italy (13%), Spain (5%) and the Netherlands (4%)[5]. Interestingly, if we compare this assertion to a research based on the geo-localiza-

---

[1] *See chapter I, page 2.*

[2] Thomas, P., et al., *Disaster in motion: 3.4 million travelers poured into US as coronavirus pandemic erupted,* ABC News, 2nd of April 2020.

[3] Brooke, J., *Growth beyond Limits: 1945 to Present,* cit.

[4] ECDC, *Timeline of ECDC's response to COVID-19,* accessed on 13th of May 2021.

[5] Pullano, G., et al., *Novel coronavirus (2019-nCoV) early-stage importation risk to Europe,* Euro Surveill., 25,4, 2020.

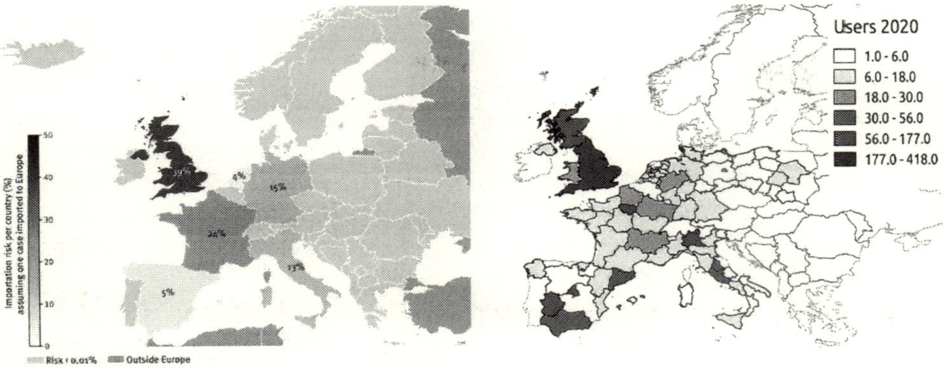

Figure 23: Left: Geo-localization of tweets containing the words "pneumonia", from: Lopreite, M., Early warnings of COVID-19 outbreaks across Europe from social media, Scientific Reports, 11, 2147, 2021; Right: probability of risk of importing Covid-19 cases depending on the number of connections between Europe and the Chinese city of Wuhan, from: Pullano, G., et al., Novel coronavirus (2019-nCoV) early-stage importation risk to Europe, Euro Surveill., 25,4, 2020.

tion of tweets containing the word "pneumonia"[6] before the official outbreak of Covid-19, we have a clue that confirms that these countries may have recorded undetected cases of the disease before its "official" arrive in Europe. The Netherlands, Germany, UK, France, Italy and Spain, where the highest number of tweets with the word "pneumonia" were recorded, are also the European countries with the largest share of import from China[7].

However, in the long run, which are the factors contributing to the spread of the virus? In May 2021, the European countries with the highest incidence of Covid19 were Czechia and Slovenia and by January 2022 the situation had not changed much[8]. Therefore, in order to find a possible explanation for the number of cases, we compared the incidence of Covid-19 cases per 100.000 inhabitants with the Globalization index (which aims to measure the number of international connections globally), the health consumer index and the population density of all the 28 MSs of the EU.

We found a possible relationship between case trends and the population density, though not a significant connection with the globalization index. Furthermore, we could hypothesise an inverse relationship between the health consumer index and the number of cases: the higher the index was, the lower the cases that were detected, though this is not entirely true, as some countries with an high index also had high levels of cases.

[6] Lopreite, M., *Early warnings of COVID-19 outbreaks across Europe from social media,* Scientific Reports, 11, 2147, 2021.

[7] Eurostat, *China-EU - international trade in goods statistics*, March 2021; Ward, M., *Statistics on UK trade with China,* House of Commons Library, 7379, 2020.

[8] Statista, *Incidence of coronavirus (COVID-19) cases in Europe as of Dec 12, 2021, by country,* 9th of May 2021.

Thus, we can conclude that political decisions may have played a decisive role. This was also underlined by the OECD, which argued that the difference in the number of incidences could be attributable to a "combination of density plus other factors such as lack of appropriate measures"[9]. We cannot make similar considerations for the Justinian plague, but the conclusion we have reached underlines the importance of political decisions in a period of crisis. Measuring the effects of political decisions during the first plague Pandemic in Early Medieval Europe is complicated by the concomitance of wars and migrations but we can nevertheless highlight some interesting points.

Procopius and other historians, indeed, reported that Justinian appointed a *Referendarius*, who "supervised the matters"[10]. In other words, we can argue that the Emperor may have wanted to centralize the management of the crisis; in fact, at the time Justinian had to cope not only with the increasing need to dig more burials, but also with the demands of workers and shortages. As we have seen in the course of this analysis, both the Justinian plague and the Covid-19 pandemic had effects on the economy, from the general level of trade to the scarcity of goods. Just to give an example of the severity of shortages, we can mention that the ongoing global shortage of microchips is forcing several factories to close and is motivated by the strains on global supply chains triggered by the pandemic; this can easily become a serious problem, as microchips are the basis of cars, household appliances, and electronic devices[11].

Moreover, both pandemics brought social problems stemming from the economic ones, which required a strong response by the state. Hence, the above-mentioned appointment of a *referendarius* in Justinian's time and the adoption of the state of emergency in contemporary Europe. However, as noted recently, pandemics and crisis can also pave the way for an authoritarian transformation of the state; this is due to the fact that crises undermine the foundations of power. This is an interesting point to which we will return when dealing with migrants and climate change. Democracies are indeed designed to cope with crises, and we can argue that Covid-19 exacerbated existing trends in Hungary and Poland, but as climate change could trigger more frequent crises in the future, will democracies endure? We should not forget that the Empire had to face not only a pandemic, but also other crises and wars; nonetheless, in the near future Europe could face new migration flows, possible climate-related conflicts, as well as the consequences of climate change in its own territory. Moreover, both pandemics and these events could be at the basis of increasing waves of social unrest: protests indeed are already occurring in these days and also occurred in Justinian's time. As we shall see in relation to migration, crisis and pandemics could lead a society to adopt an inward-looking approach, closing to foreigners and to the outside world: this is what we saw, for instance,in chapter one when we noted a general "fortification" of Europe's settlements. This strengthening of defences, probably triggered both by the migratory movement and the realisation that the Justinian plague was coming from outside,

[9] OECD, *The territorial impact of COVID-19: Managing the crisis across levels of government,* 2020.
[10] John of Ephesus in Ps. Dyonisius of Tel-Mahre, Chronicle, Part III, cit.
[11] Il Post, *La crisi dei microchip è sempre più grave,* 5th of April 2021.

may have led to a reorganization of the way of living. This seems to be proved by the abandonment of the *villae* and the sudden appearance of hillforts and may have led to a reorganization of society. Again, we will deal with this issue when we focus on migrations, but we can plausibly trace the origins of feudalism, which started to develop in the fifth and sixth century[12], also to the effects of the plague: the increase in bands of warriors protecting prominent men may have become necessary due to the general uncertainty pro voked by the plague's effect. Paul the Deacon, for instance, reports of deserted settlements and roads[13] and although many people died, we are also told that many others sought refuge in other settlements or migrated, probably wandering across Europe.

Poverty and the risk of poverty are another problem common to both periods. Already before the pandemic, 21% of the inhabitants of the European Union were at risk poverty[14] and Covid-19 exacerbated the situation, as we have seen in the course of this research. In fact, the most fragile people, both economically and in terms of health are those who pay the highest price in times of crisis. According to the World Bank, the Covid-19 pandemic will drive a further 124 million people in poverty, 80% of which in middle income countries[15]. This means that more and more people will experience difficulties not only with household expenses, but also with mortgages or rents[16]: as for the USA, for instance, it was estimated that between 30 and 40 million people could have lost their house by the end of 2020[17].

Moreover, this situation is affecting a society that has not yet fully recovered from the economic crisis. Not surprisingly, it is becoming a common trend – for now mainly in the USA – to consider alternative forms of living such as vehicles; as Angus Duff and Scott Ranking, from the School of Economics of Thompson River University, argue, increasing housing costs and precarious employment are leading people to leave expensive urban centres and return to live with their parents, buying or renting tiny apartments and building tiny homes[18]. We could therefore be at the eve of a reorganization of spaces and living, as it also happened in Early Medieval Europe. In this sense, the phenomenon of the "Great resignation" is extremely interesting: tired of the pace of work (and frequent burnout), more and more workers are abandoning their jobs, taking advantage of retirement incentives, or turning to other types of work, in what is already being called the 'Great Reshuffle'. We can, to some extent, compare this phenomenon to the abandonment of the fields we read about in the accounts historians of early medieval Europe. Faced with a phenomenon as historic as the pandemic, people reorient their lives, deciding to aban-

---

[12] Ganshof, F., *Feudalism,* University of Toronto Press, Medieval Academy of America, 1996.

[13] *See chapter I, page 17.*

[14] Eurostat, *Over 20% of EU population at risk of poverty or social exclusion in 2019,* cit.

[15] Yonzan, N. et al., Bank, *Available data and estimates of the impact of the COVID-19 pandemic on global poverty,* May 2021.

[16] Center on Budget and policy priorities, *Tracking the COVID-19 Recession's Effects on Food, Housing, and Employment Hardships,* 7th of May 2021.

[17] National Low Income Housing Coalition, *30-40 Million People in America Could Be Evicted from Their Homes by the End of 2020,* 7th of August 2020.

[18] Duff, A., Rankin, S., *Exploring flexible home arrangements – an interview study of workers who live in vans,* Career Development International, 25, 7, 747-761.

don work out of fatigue, boredom or disillusionment. In an age of possibilities such as the one in which we live, we could therefore witness, in the medium term, epochal changes in the labour market, with all that this entails.

Another trend common to the two periods is the demographic decline: it is estimated that between the sixth century and 650 CE, European population fell from 27.5 million to 18 million[19]. This is mainly reflected in the abandonment of fields, but also in the decline of trade, with its consequent economic effects. Similarly, Europe is facing a declining birth rate, which could have important consequences for the economy: according to Bloom et al., if the fertility rate remains at its current level, there will be a decline in the working age population and lower *per capita* income[20]. Even though this trend was already evident before the pandemic,the current situation could adversely affect this decline.

From a sociological point of view, both pandemics had important impacts on individuals. Important social and cultural centres such as "theatres, hippodromes, and circuses [...] were all discontinued": interestingly, this is not a contemporary description, but the account of the historian Procopius, who also describes phenomena of mass hysteria and vandalization of the cities. As previously mentioned, several contemporary researches are focusing on the impact of schools and community centres closures, and evaluating the effects that this may have on both health and future employment prospects as well as on early school leaving.

Moreover, episodes of mass hysteria and scapegoating linked to the pandemic, as also described in ancient sources, have been observed even in our times. We can describe them as the natural reaction of a population that is forced to abandon its habits from one day to another. However, in such a globalized world, we are faced with the possibility that pandemics will become more frequent, again imposing lockdowns, social distancing, and restrictive measures that impact on social habits. If it will be so, will social cohesion vane?

It is often said that we will have to learn to live in a more digital world, working, studying, and meeting each other via web. However, from a sociological and philosophical point of view, might this be the solution? The cognitive effects of social distancing on the children deprived of schooling seem to confirm that a further digitalization of schooling could be detrimental to health[21]. Moreover, from a more practical point of view, this could exacerbate the digital divide, by further excluding those who do not have access to internet, as was also seen during local lockdowns. Furthermore, from an economic point of view, it is true that online shopping is developing, and retailers are mainly integrating their shops with online versions of their stores, but this process is also impacting on small producers, who find it difficult to give visibility to their goods[22].

[19] Josiah C., *Population in Europe* in: Carlo M. Cipolla, ed., The Fontana Economic History of Europe, Vol. I: The Middle Ages, 25-71, 1972.

[20] Bloom, D., et al., *The cost of low fertility in Europe,* National Bureau of Economic research, 2009.

[21] Lopez-Bueno, R., et al., *Potential health-related behaviors for pre-school and school-aged children during COVID-19 lockdown: A narrative review,* Preventive Medicine, 143, 2021.

[22] Melisande, C., et al., *The macro-economic impact of e-commerce in the EU Digital Single Market,* European Commission, 2015.

Pandemics, therefore, could change our habits, but we should also consider *how* we want our habits to be changed. Importantly, indeed, pandemics also affect our intimate habits such as people's relationship of men with dead. This is also evident from the descriptions in ancient sources of mass graves and the need to find burials for dead people: in the contemporary world, we cannot forget the images of trucks carrying coffins in Bergamo[23], mass graves in New York[24] and the huge number of pyres in India[25]. The impact that these images have had on society could be similar to that on Procopius's contemporaries: as the historian reports, when the smell of buried corpses reached the city, it "distressed the inhabitants still more", thus probably alluding to a common sense of distress that is also evident from the narration of Pseudo-Dionysius, who tells of a man carrying a tablet with its name in case he dies[26]. This description sounds extremely topical and resembles those people who died in hospitals, far from their relatives, and unable to see them due to the emergency situation. Moreover, both pandemics also affected relationships among people: the ancient historians tell of people abandoning their factions to help the sick and the fragile ones, but also to people abandoning their relatives. Again, during the Covid-19 pandemic we saw a common sense of solidarity and even NGOs claim that more people were willing to help than before the pandemic, although we have saw growing episodes of violence, scapegoating and hate speech. These events, therefore, have had and will have a huge impact on society: such a change in burial habits as we witnessed in the post-pandemic Europe during the Migration Period, indeed, seem to reveal an important change in the societal practices.

The crises triggered by pandemics or epidemics also urge men to concentrate their efforts on health care, looking for treatment and remedies for the disease. In Chapter I we saw that the development of hospitals and health centres might be related to the Justinian plague; at the same time, contemporary investment in the health care by European authorities could be seen, in the future, as driven by the necessity to face future challenges to our health systems as the one posed by Covid-19. Nonetheless, when faced with a new and mysterious disease, men could also decide to rely on "unofficial" remedies such as magic spells at the time of Justinian or other composites in our time[27]. This choice may also be prompted by mistrust in the official treatment which, as we could argue, reflects the mistrust in author ities that we have described above, as also stated by psychologists Palamenghi et al.[28]

A pandemic could also have geopolitical implications, as evidenced nowadays by the importance of vaccines – which are becoming a "soft power tool" – as well as by the tensions over vaccine imports and exports, and the fake news associated

---

[23] Il Post, *Le bare dei morti di COVID-19 di Bergamo portate sui camion dell'esercito*, 19th of March 2020.

[24] Hennigan, W., *Lost in the Pandemic: Inside New York City's Mass Graveyard on Hart Island*, 18th November 2020.

[25] BBC, *India Covid: Delhi builds makeshift funeral pyres as deaths climb*, 27th of April 2021.

[26] Pseudo-Dionysius of 'Tell-Mahre', *Chronicles*, cit.

[27] Slattery, G., *Aspirin in honey: dubious COVID-19 'cures' spread in Brazil*, Reuters, 17th of June 2020.

[28] Palamenghi, L., et al., *Mistrust in biomedical research and vaccine hesitancy: the forefront challenge in the battle against COVID-19 in Italy*, Eur J Epidemiol 35, 785-788, 2020.

to the serum, the authors of which are considered China and Russia. In the Middle Ages, on the other hand, the demographic decay triggered by the plague would have pushed the Slavs to move into the Balkans, thus changing the *status quo* of the peninsula; moreover, according to some authors, the difficulties encountered by the Eastern Roman Empire might have also been at the basis of the Arabs' expansion.

Notably, the perceived shortage of labour in Europe is leading many African migrants to leave their home: the pandemic has thus become a *pull factor* but it is also a *push factor* in several African countries, where 20 million people lost their job. It is important to stress that this is almost the same amount of people who migrated within Africa in the last ten years[29]. Another migration-related problem is the reduction in remittances, which are projected to drop by 23%, since migrants are losing their jobs in foreign countries; remittances proved to be effective in contrasting poverty and child labour in low and middle-income countries and even surpassed FDI in 2019 in the entire African continent. Such a sharp reduction could lead more people to leave their countries looking for jobs: again, we should also consider, however, that pandemics could become more frequent in the future and that climate changes might also trigger more conflicts and migration, thus complicating the picture. It is thus time to focus on migration in both periods.

3.1.2 Across, into, and again across Europe?

Migrations, as we have seen, can have multiple causes; that is why we commonly distinguish between *push* and *pull* factors. This research has shown that among the push factors we can mainly list conflicts and economic reasons; we have also found that climate change may be among the causes of contemporary waves of migration, but it may also have been behind the movements of Germanic tribes. This is not only because climate change impacts on the livelihoods of farmers and, in general, on the whole economy. Climate change, indeed, could also fuel violence and economic problems, triggering disputes over lands that may degenerate in larger conflicts between opposite social groups such as shepherds and farmers. This section, therefore, will be devoted to comparing the migration waves in both the Migration period and our era, with the aim of identifying common trends and causes.

We will start our analysis from the Migration period. Importantly, by comparing Cesar's historical account with that of Tacitus, we understood that something was already changing within the Germanic society in the first centuries of the first millennium: new organizations and new confederations, such as the Alemanni, the Thuringi, the Franks and the Saxons were emerging in Barbaricum. As we have seen, historians argue that this was both due to an ongoing process of demographic expansion and to the attraction exercised by the neighbouring Roman Empire, which, despite the existence of a *limes,* had important contacts with the Germanic tribes. This "revolutionary process" deeply interested the Romans, who decided to support certain princes and often co-opted some of them as they acknowledged that a transforma-

[29] African Union, *Africa labour migration outlook in the post Covid-19 era*, Policy Brief I, 2020.

Figure 24: Barbarians leaving Germania for the Empire, as represented in the "Plomb of Lyon" (297 CE); Mathisen, R., Peregrini, Barbari, and Cives Romani: Concepts of Citizenship and the Legal Identity of Barbarians in the Later Roman Empire, The American Historical Review, Vol. 111, No. 4 (October 2006), pp. 1011-1040, 2006

tion from a kingship to an oligarchy was taking place[30]. However, in the following years, a migratory movement started: a limited number of people entered the Empire, ending up as far as in Africa (the Franks), while others decided to become *foederati,* that is to say to settle with their families and fight for the Romans. We will later try to explain these movements. Importantly, this process of social transformation was taking place at the same time as the integration process of the Germans into the Roman territory: in 212, indeed, Caracalla issued the *Constitutio Antoniana*, which granted the Roman citizenship to all the inhabitants of the Empire: among them, there were hundreds of thousands of Germanic descendants who had been settled within the Empire to cultivate the deserted lands and pay tribute.

These descendants, in fact, were not a small number: the process of resettlement within the Empire had started with August, who had settled 50,000 Goths on the Danube; then Tiberius (14-37) moved 40,000 Germans into Gaul; Nero (54-68) gave lands in the Balkans to more than 100,000 people from across the Danube; Marcus Aurelius settled the Quadi, Vandals, Iazyges, Naristae, and Marcomanni in the Empire; under Claudius, the defeated Goths (270) became farmers in Roman lands; Constantius Chlorus (293-306) assigned deserted lands to the Barbarians too;

[30] Wolfram, H., *The Roman Empire and Its Germanic Peoples,* University of California Press, 1997.

Constantine I (303-337) settled the Franks in the deserted lands in Gaul and made them *foederati,* while also giving lands in Thrace, Scythia, Macedonia and Italy to more than 300,000 Sarmatians, followed by Ausonius in 368; moreover, between the 360s and the 370s, part of the Alamanni were relocated in Gaul and Italy, where this population started farming along the Po; in 377, they were followed by the Goths and Taifals, who became cultivators in Rhegium and Parma; in 386, "the nation of Greuthungi" was brought into the Roman Soil" and twenty years later a group of Scirians became *coloni* in the Eastern Empire[31].

Looking at he so-called Barbarian invasions with a contemporary perspective allows us to see them in a different light: more than invasions, in fact, these movements seem to have been migrations of peoples who followed the same routes as their ancestors who had settled in the Empire centuries before, and with whose descendants they had probably remained in contact. Further studies could therefore investigate the effects that the settlement policies had on the migrations of Germanic tribes into the Roman Empire, but we can already argue that they certainly provided an incentive for Germanic tribes to move into the Empire, probably even more so after 212, when the descendants of these settlers became Roman citizens, with all that this entailed.

Moreover, in *Barbaricum*, Germanic chieftains were highly influenced by Rome, not only from a political point of view, but also from a cultural perspective: the Empire could offer them richness, prestige, and power, something that tribal life could not[32]. Moreover, the gifts provided by Rome and the proximity of the Imperial power had already "raised the stakes in barbarian politics", as previously said[33]. Social status and political authority were manifested through the use of Roman stylistic motifs and Roman goods, be they gifts or the result of trade or raids; this makes the historian Guy Hansall claim that the Germanic tribes could have been more bellicose among themselves than against Rome, on which their status depended[34]. It is interesting to note what the historian says: "what would happen to such groups if the Empire ceased to exist as a valid model for legitimate rule? These units could be destabilised and, when they were, the threat to the security of the frontiers was that much greater"[35].

On the other hand, Rome was also becoming weaker, due to a series of factors we have not dealt with. Therefore, a combination of issues, such as the *attraction* of Rome, coupled with its political weakness, could have provided the necessary incentives for an increasing number of Germanic tribes to cross the borders.

Nonetheless, before going on with the Migration Period, we must focus again on our time: indeed, what can we grasp from the analysis of the Migration Period? In our opinion, although we are dealing with two extremely different periods from both a technological and societal point of view, we can nonetheless find some interesting similarities.

---

[31] Mathisen, R., Peregrini, Barbari, and Cives Romani: Concepts of Citizenship and the Legal Identity of Barbarians in the Later Roman Empire, cit.

[32] Pohl W., *The Avars: A Steppe Empire in Central Europe, 567–822,* cit.

[33] Halsall, G., *Barbarian Migrations, and the Roman West, 376–568,* cit.

[34] *Ibidem.*

[35] *Ibidem.*

We have already stated that the aim of this research is to find similarities and common trends between the two historical eras; therefore, it is necessary to adopt a *longue durée* approach, not focusing only on the twenty-first century, but also on the previous decades. As nowadays we are dealing with migratory waves coming mainly from Africa and Asia, we also have to stress that, until the Decolonization process, these territories were governed by the same States to which now migrants are heading to: the European ones. Interestingly, European states governed these territories with the same approach of the Roman Empire: they appointed chieftains favourable to them while also supporting certain tribes or ethnicities, thus creating the basis for the subsequent ethnic tensions such as the ones experienced in Rwanda and Burundi[36].

Then, in the aftermath of the Second world war, new political instances would lead to the decolonization process and to the formation of independent states. However, the states that came out from this process where extremely fragile, since colonial rulers had shaped them regardless of the previous societal organization: the administrative reorganization under the European powers, indeed, had transformed the relationship between groups favouring some of them. As African History professor Gentili claims, the administrative rationalization of the territories into defined and territorial entities meant a remodelling, a reconstruction or even the creation of the traditional authority, with the contemporary opposition and collaboration of local chieftains[37]. Thus, on the eve of the decolonization, united fronts emerged to achieve independence; however, once self-government had been achieved, this unity, represented in the single (independence) party, became the norm, and the lack of democracy was justified by the need to rebuild the nation[38]. Then, from the mid-1960s onwards, a series of military coups transformed African states once again; moreover, authoritarian governments – not only in Africa, but in other parts of the world as well – were also linked to the Cold War divisions: the former colonial powers therefore continued to play an important role in these areas. In addition, many of these post-independence rulers had studied in the former colonising states or had served in the colonial armies. With the end of the Cold War came the first democratic elections, but internal tensions did not stop, and a growing number of African countries are now experiencing democratic backsliding, while the continent is ever more at the centre of geopolitical disputes[39] for its abundance of resources and for being a possible outlet market for goods and services[40].

Returning to the comparison with the Migration period, we can observe some similarities between the two periods in the relation between the two "worlds". Although we cannot describe the Roman influence on the Germanic tribes as "colonisation", we can nonetheless state that the Empire exerted a considerable influence on *Barbaricum*. The appointment of certain rulers and the support given to specific chieftains and tribes were indeed useful to maintain these lands under con-

---

[36] Gentili, A., *Il leone e il cacciatore,* Carocci Editore, Roma, 2015.

[37] *Ibidem.*

[38] *Ibidem.*

[39] Cheeseman, N., *A divided Continent,* BTI, 2018.

[40] Alden, C., *Emerging Powers and Africa: From Development to Geopolitics,* IAI, 19, 2019.

ditions favourable to the Roman Empire; at the same time, the Germanic tribes derived their status from the proximity of the Empire and probably they underwent a process of federation also because of the influence and the attraction of the neighbouring power. These once-divided tribes, therefore, organized themselves in a societal framework influenced by the dominant power, probably with the aim to emulate it, given the appeal of the Empire.

In more recent times, post-colonial African states, in particular, were organised under the same administrative and political structure of the former colonising states, despite their traditional divisions and boundaries: this "process, inspired by the belief that the western state was the model to be followed, created tensions in society, as mentioned. These tensions, in turn, triggered the authoritarian response by the army, bringing to power military governments which had also important ties with the former colonial states.

Interestingly, we could state that the above-mentioned early migrations, such as those of the Franks and of the Heruls, might have been provoked by similar tensions in *Barbaricum* following the development of new political entities. These tribes could have decided to leave at the beginning of this 'centralisation' process, as some Norsemen would later do during the formation of the first Norwegian unitary state.

Moreover, just like modern military leaders had studied in the West, Germanic leaders were also trained in the Empire since the Romans used to take the sons of Germanic nobles as hostages[41], in order to speed up the process of Romanisation.

In addition, in the light of this new information, we could claim that the movements of Germanic tribes could have also been motivated by the desire to achieve the same cultural results and prestige of the Roman Empire. The Barbarian invaders, indeed, certainly did not want to destroy the Roman cultural heritage, as it is evident from the adoption of the Roman laws and from the perpetuation of the Roman culture, albeit with the inclusion of Germanic elements. At the same time, the interviews of contemporary migrants highlight that they are moving to Europe because of its attraction in terms of wealth, democracy, and higher standards of living. Something which is lacking in their countries of origin.

We also argued that the attraction exercised by the Roman wealth and culture could not constitute a sufficient stimulus for people to move, in the same way as – we may argue – "poverty" cannot be considered a necessary motivation to migrate, as mentioned in chapter two. Violence and conflict, indeed, are generally the other main drivers of outmigration. And we have already seen that these tribes were probably leaving their lands because of the federation process and the arrival of the Huns.

The Huns first created a confederation of tribes over which they ruled and allied themselves with the Romans; then, after the death of Attila, the German tribes revolted and brought about the end of the Hun Empire. Now they were free to move into the Roman Empire, also driven by another variable: climate change.

Notably, we can say that the same is happening now: European states have already signed several agreements with North African countries and Turkey to stop mi-

[41] Lee, D., *The Role of Hostages in Roman Diplomacy with Sasanian Persia*, Historia: Zeitschrift für Alte Geschichte, 40, 366-374, 1991.

gratory flows[42], giving them great bargaining power. These agreements have effectively reduced the number of arrivals in 2016/17 at the expense, however, of a dismissal of European humanitarian responsibility, and with the result of "outsourcing" asylum management to countries that do not recognize the right to asylum[43].

However, one of the most important lessons of the Migration Period is precisely that migration cannot be stopped. The Hun empire did indeed delay a migration process that had already started in the 3rd and 4th centuries, but it did not stop it.

As we are already seeing, migratory flows are increasing again. Indeed, in the summer of 2021 Spain faced a migration crisis in Ceuta and Melilla as Morocco was accused of doing nothing to prevent migrants from arriving in the two enclaves: in particular, several commentators claimed that behind the crisis there could be the Spanish decision to host the leader of the Polisario Front, Brahim Ghali, who is fighting for the sovereignty of Western Sahara against the Moroccan claims[44]. At the same time, Italy asked European authorities to step up assistance to Libya when, according to its intelligence, 70,000 migrants were ready to cross the Mediterranean. Moreover, EU authorities are also negotiating the renewal of the EU-Turkey agreement[45]. These recent events, therefore, seem to confirm what we have argued before, namely that migratory pressure will not stop but rather is bound to increase. On the other hand, negotiating agreements to stop the migratory flows are not a solution: in fact, if projections are correct and climate refugees will indeed be between 25 millions and 1 billion in the near future[46], agreements or borders will not prevent this large human flow from moving. We will deal with this issue in the next paragraph.

Another interesting theme of comparison is the political transformation of Europe: following the Migration Period, new kingdoms were formed: these "Roman-Barbarian" entities were mixed entities, where, for instance, Roman laws continued to apply, as mentioned above. Moreover, the Visigoths developed their own culture only when they settled in present-day France and not before; furthermore, Ralph Mathisen, author of the book "Romans, Barbarians, and the Transformation of the Roman World" argues that the very elites of the Germanic tribes were "international", in the sense that they often stayed in distant courts and had blurred identities[47]. This is not entirely relevant for our analysis, but it gives us the idea of a general cultural reshaping of Europe, which integrated several distant elements. What could interest us, however, is the fact that the political organization changed in Early

---

[42] Terry, K., *The EU-Turkey Deal, Five Years On: A Frayed and Controversial but Enduring Blueprint*, cit.

[43] Palm, A., *The Italy-Libya Memorandum of Understanding: The baseline of a policy approach aimed at closing all doors to Europe?* cit.

[44] Hedgecoe, G., *Migrants reach Spain's Ceuta enclave in record numbers*, BBC, 18th of May 2021; Nicastro, A., Ghali, il leader del Sahara diventa un caso diplomatico tra Marocco e Spagna, Corriere della Sera, 19th of May 2021.

[45] ANSA, *Migranti: Johansson in Turchia per discutere accordo con Ue*, 6th of May 2021Bassetti, F., *Environmental Migrants: Up to 1 Billion by 2050, Foresights* – The CMCC observatory on climate policies and futures, 22nd May 2019.

[46] Bassetti, F., *Environmental Migrants: Up to 1 Billion by 2050, Foresights* – The CMCC observatory on climate policies and futures, 22nd May 2019.

[47] Romans, *Barbarians, and the Transformation of the Roman World*, edited by Mathisen, R. and Shanzer, D., Routledge, 2016.

Medieval Europe with the rise of feudalism, which many studies trace back to a Germanic custom described by Tacitus, the *"comitatus"*[48], which started to be applied to the new states in the post-Roman world.

As Europe will experience new migrations in the future, its political organization is an issue that we will have to address: how will European society evolve following the arrival of new migratory flows? Caracalla, for instance, enacted the citizenship reform, paving the way for the subsequent transformation of the Empire; will European countries change their citizenship laws? Interestingly, this is a much-debated topic nowadays.

What is more, in such a globalised world faced with increasing transnational phenomena such as migration and pandemics, how can our model of the state evolve? In the face of crises that will require an increasingly rapid response, what model should we follow? And above all, will climate change lead to new intra-European migrations?

3.1.3 Are we living in another Migration Period? Problematic issues

The aim of this study was also to discuss the possibility that we might experience historical processes similar to those of the Early Medieval Ages' Europeans. The *Migration period*, indeed, was not only a period of migrations, but also of pandemics and cli mate change. It was, therefore, a period dense of social changes and crisis. Moreover, Europe was highly influenced from that period: new kingdoms were formed, new social structures emerged, and new people arrived. Importantly, these huge changes were triggered by the simultaneous presence of pandemics, migrations and climate change, which have interacted with each other, influencing historical development. Indeed, we can only speculate on what would have happened if global cooling had not occurred: would have the German tribes remained in *Barbaricum*? Would have the Huns arrived anyway? And what if the plague had not outbroken? Would have the Slavs migrated into the Balkans? We can already grasp the relevant historical consequences if things had gone differently. Importantly, nowadays Europe is at a crossroad, since it is experiencing the same crisis that took place in the Migration period.

Since climate change may worsen and cause more damage in the future, we must analyse its possible future effects from an holistic perspective. On the basis of the diagram we drew in the second chapter, we can see that the anthropogenic climate change could cause droughts, floods, and pandemics. Importantly, these factors can cause economic losses and food insecurity, which are the main drivers of conflict and migrations. Lastly, we argued that migration and conflicts cause displacement and a progressive flow into cities and urban areas, which in turn grow haphazardly, thus causing disruption of habitat. This creates the conditions for other pandemics both because of the creation of densely populated areas, but also because of the possibility of a zoonotic spill-over.

However, we did not include in the analysis other factors such as geopolitical issues, wars and conflicts caused by other drivers that could still have an impact on ex-

[48] Ganshof, F., *Feudalism*, cit.

isting problems caused by climate change or exacerbate them further. We argue, in fact, that climate change sets the conditions for conflicts, migrations and pandemics, but we cannot determine whether these will actually take place or interact with existing tensions. Indeed, we agree with researchers who claim that climate change is a "threat multiplier", because it aggravates "existing socioeconomic stresses"[49]. At the same time, however, we claim that in the next future climate change may become – at least indirectly – the main cause of migrations and conflicts, which could also be exacerbated by more frequent pandemics. Therefore, how will our societies react to these changes?

We have seen that both in the Migration Period and in our era, pandemics also created the conditions for major geopolitical changes: the Slavs entered the – presumably – abandoned Balkans, while nowadays migrants are heading Europe because of the economic difficulties in their countries but also due to a perceived shortage of labour in Europe. However, as these events may become more frequent in the future and interact further with each other, the presence of simultaneous pandemics and migration crises could be detrimental to states, not only from an economic but also from a social point of view: in recent years, indeed, we saw that several states decided to close their borders in response to the migratory crisis. Interestingly, this was the same response that was adopted, for sanitary reasons, by the great majority of the world's countries. We may also contend that European states like Hungary or Poland were already experiencing democratic backslidings, but both the refugee crisis and the Covid-19 one gave them the opportunity to further the authoritarian descent. However, importantly, even a still democratic State such as the United Kingdom decided to leave the European Union in the same period; interestingly, according to several analysis, immigration "was one of the single strongest issue driving Brits to vote Leave"[50]. Moreover, during the "refugee crisis", Germany, Austria, Denmark, Sweden and France reintroduced checks at internal borders, thus triggering a "Schengen crisis"[51]: importantly, the abolition of the internal border controls is widely considered one of the pillars of the EU. Furthermore, at the very eve of the Covid-19 pandemic, the European solidarity faltered as France and Germany blocked the export of masks[52], while the European border closures were motivated by sanitary reasons[53].

Therefore, will further migratory waves and pandemics create the conditions for a "more closed" world and for the disintegration of the European Union? On the one hand, it is important to stress that the European Commission pressured both

[49] Huntjiens, P., Nachbar, K., *Climate Change as a Threat Multiplier for Human Disaster and Conflict*, Working paper 9, The Hague Insititute for Global Justice, 2015.

[50] Garret, A., *The Refugee Crisis, Brexit, and the Reframing of Immigration in Britain*, Europe Now, Council for European Studies, 1st of August 2019.

[51] Colombeau, S., *Crisis of Schengen? The effect of two 'migrant crises' (2011 and 2015) on the free movement of people at an internal Schengen border*, Journal of Ethnic and Migration Studies, 46, 11, 2258-2274, 2020.

[52] European Parliament, *German ban on medical equipment exports, 16th of March 2020; European Parliament, Masks intended for Italy blocked by France*, 3rd of April 2020.

[53] European Commission, *Temporary Reintroduction of Border Control*, accessed on the 20th of May 2021.

TABLE 1
MANUSCRIPT PRODUCTION IN ABSOLUTE NUMBERS PER CENTURY
(sixth to fifteenth centuries)

Area	*Sixth*	*Seventh*	*Eighth*	*Ninth*	*Tenth*
Central Europe[a]	0	0	0	0	0
Bohemia	0	0	0	0	0
British Isles[b]	81	1,026	5,474	7,926	9,793
France	1,682	2,441	15,920	74,190	12,752
Belgium	0	127	1,111	3,029	1,555
Netherlands	0	26	60	82	58
Germany	0	0	7,503	59,771	45,703
Switzerland	0	30	594	5,330	1,799
Austria	0	0	2,735	9,414	0
Italy	10,194	4,478	6,536	20,307	15,215
Iberia[c]	1,594	2,512	3,770	21,693	48,763
Western Europe	13,552	10,639	43,702	201,742	135,637
Increase per century (percent)		–21	311	362	–33
	Eleventh	*Twelfth*	*Thirteenth*	*Fourteenth*	*Fifteenth*
Central Europe[a]	3,983	27,530	120,987	301,833	376,650
Bohemia	657	1,136	5,377	42,066	45,363
British Isles[b]	20,360	81,044	200,654	155,513	208,729
France	45,061	197,831	510,828	564,624	1,195,783
Belgium	8,529	43,219	119,588	106,148	572,124
Netherlands	354	1,731	2,066	13,179	171,974
Germany	49,548	166,876	270,392	293,814	515,116
Switzerland	1,090	2,355	3,821	6,349	10,652
Austria	2,808	37,370	37,408	39,777	88,623
Italy	38,768	95,207	253,013	879,364	1,423,668
Iberia[c]	40,871	114,422	237,818	344,284	390,478
Western Europe	212,030	768,721	1,761,951	2,746,951	4,999,161
Increase per century (percent)	56	263	129	56	82

Figure 25: Manuscript production from the Early Medieval Period to the end of the Middle Ages: from Buring, E., Van Zanden, J., Charting the "Rise of the West": Manuscripts and Printed Books in Europe, a LongTerm Perspective from the Sixth through Eighteenth Centuries, The Journal of Economic History, Jun., 69,2, pp.409-445, 2009.

France and Germany to lift the blockade[54] and then Europe moved towards a general solidarity during the pandemic[55]. However, on the other hand, a general agreement on the redistribution of migrants has not been found yet[56] despite the reform of the Dublin Regulation in September 2020[57]. These examples clearly show that these combined crises could divide the European states; however, we have to consider that increasingly frequent crisis will also require rapid and strong responses. For instance, Mastrojeni and Pasini claim that the melting of the Himalaya glaciers due to global warming could trigger a crisis which in turn might escalate in a major war between countries like China, India, Pakistan and Russia, all of which, importantly, detain the atomic bomb[58].

Indeed, it is crucial to remember that as the world is increasingly interconnected, local processes have the potential to have an impact on distant areas. While, for instance, climate change may have displaced the Avars from their Asian homeland, in that case the effects on Europe were not visible until two hundred years later; on the other hand, nowadays, deteriorating climatic conditions in Syria have created the conditions for a local conflict that is impacting on the whole world, and has (nearly) led to an escalation of tensions between Russia and the USA[59]. What would Europe do in the face of such a crisis?

Mastrojeni and Pasini argue that European states could respond to the growing migratory waves with new agreements aimed at curbing the number of arrivals, thus also indirectly favouring authoritarian drifts, and a general decrease in the states' respect for human rights[60].

[54] Ministero dello Sviluppo Economico, *Sblocco delle esportazioni di dispositivi sanitari da Germania e Francia*, 15th of March 2020

[55] European Commission, *Coronavirus: European Solidarity in action*, accessed on the 20th of May 2021.

[56] ISPI, *Sui migranti l'Europa non c'è*, 12th of May 2021.

[57] European Parliament Think Tank, *Reform of the Dublin system*, 30th of September 2020.

[58] Mastrojeni, G., Pasini, A., *Effetto serra, effetto guerra*, cit.

[59] Yacoubian, M., *What is Russia's Endgame in Syria?* United States Institute of Peace, 16th of February 2021.

[60] Mastrojeni, G., Pasini, A., *Effetto serra, effetto guerra*, cit.

However, external pressures can also trigger populist responses in European states, as we have seen in recent years with the growth of nationalist and populist parties in most European countries[61]; could the growing global challenges such as pandemics and migrations crises, therefore lead to a crisis in the Western democracies, as already happened in several countries[62]?

On the other hand, even if the European rule of law succeeds in resisting to the democratic decay triggered by the rise of populist parties, Europe may find itself isolated in an increasingly 'closed' and authoritarian world; the worsening climate, in fact, could trigger tensions and wars not only within states, but also between them, as the example of the melting of the Himalayas underlines. Possible tensions could arise, for example, over the use of the Jordan River or the sharing of water in the Nile basin. We have already seen that most of the world's refugees are hosted in these two areas: what would happen if tensions increased in these regions? In particular, Syria and South Sudan alone are already the countries of origin of almost 9 million refugees.

In this sense, it should also be noted that neither Europe nor the USA currently host the majority of refugees: 85% of the refugees, indeed, are hosted in developing countries[63].

Nowadays, Turkey is the country with the largest number of refugees (3.3 million), followed by Colombia (1.8 million), Pakistan (1.4 million), Uganda (1.4) and Germany (1.1 million)[64]. It is also important to remember that the African population is expected to double by 2050, reaching 2.2 billion inhabitants or more[65] and so will do the Middle East, reaching 723 million people[66]. However, since Africa, the Middle East and South America are among the regions that will experience the worst weather conditions, we can assume that migratory flows will converge north, were better conditions are expected.

Global warming, moreover, could trigger more crop failures, more tensions, more conflicts and thus more migration. In the worst case scenario, the Sahel region could become almost uninhabitable and vital crops such as maize, beans, bananas and millet would decline by 35%, thus forcing millions of people to leave. Already today, in 2020 alone, according to a report published by the Internal Displacement Monitoring Centre of the Norwegian Refugee Council, there were 30.7 million climate-related internally displaced persons (IDPs) worldwide, which is three times the total number of conflict-related IDPs (9.8 million). Most of them were in China (more than 5 million), Philippines (4.449.000), Bangladesh (4.443.000), and India (3.856,000)[67]. We are already seeing the effects of climate change on migration.

---

[61] De Torre, C., *Routledge Handbook of Global Populism*, Taylor and Francis, ebook, 2018.

[62] Vachudova, M., *Ethnopopulism and democratic backsliding in Central Europe*, East European Politics, 36, 3, 2020

[63] UNHCR, *Figures at a Glance*, 18th June 2020.

[64] *Ibidem*.

[65] Ezeh, A., et al., *Why sub-Saharan Africa might exceed its projected population size by 2100*, 396, 10258, 1131-1133, 2020.

[66] UN DESA, *Total population across the Middle East and North Africa from 1990 to 2050 (in million)* [Graph]. In Statista. Retrieved May 18, 2021.

[67] IDMC, *Global Report on Internal Displacement 2021*, accessed on 27th of May 2021.

Interestingly, the University of Notre Dame, (Indiana, USA) created an index, the ND-GAIN (Notre-Dame Global Adaptation Index), which aims to measure a country's vulnerability to climate change in combination with its readiness to improve resilience. When measuring vulnerability, the authors took in account the exposure, the sensitivity, and the adaptive capacity, while when measuring readiness, they considered economic, social and governance readiness.

Moreover, the authors also classified the countries in a Vulnerability/ Readiness matrix, dividing the well-prepared states with few challenges to face, from the ones with more urgency to act and with the greatest challenges to cope[68]. Among the worst prepared and with the highest number of challenges we find countries from which large numbers of migrants already depart: Chad, Somalia, Central African Republic, Afghanistan, Eritrea, Sudan and Niger, among others. As for countries with good performance and limited challenges, we find Northern European countries such as Norway, Finland, Denmark and Sweden, but also Austria, Iceland and Germany[69].

Lastly, we must also consider the effects of Climate change in Europe. As previously stated, droughts and wildfires will become more frequent in our continent as a result of rising temperatures, and ecosystems may not be able to supply the current amount of water, which will also have major effects on agriculture and livestock. Rising sea level, flooding, agricultural losses, and water shortages will have a particular impact especially on Mediterranean regions, i.e. those most exposed to migration flows. It is important to underline that Southern and Eastern Europeans are those who are already experiencing the effects of climate change on their lives the most and those who say they might leave their countries because of global warming[70]. In the face of worsening climatic conditions and increasing waves of migration, how will Southern European countries react? Are we going to see more intra-European migrations?

In conclusion, we can argue that the Migration period changed Europe completely: the simultaneous appearance of the Justinian plague and its outbreaks together with the migrations of people and related conflicts contributed to the dissolution of the Roman Empire and the emergence of several Romano-Barbaric Kingdoms. Moreover, although we have previously argued that these kingdoms perpetuated classical knowledge, this was a limited activity; the general decay in the dissemination of culture could for example be measured for instance by the production of manuscripts, which is extremely low compared to the eight century (see figure 2). Interestingly, we set the end of the Migration Period between the eight and the ninth century, when the Justinian plague mysteriously disappeared, temperature began to increase and so did settlements, as seen in chapter one.

It is important to emphasise that the loss of culture in the Middle Ages was due to other causes and that we are not claiming that we are on the eve of a general loss of culture; however, it is also important to stress that today we are witnessing a growing climate of distrust in science and the scientific authorities. Whether this is

[68] Chen, C. et al., *University of Notre Dame Global Adaptation Index Country Index Technical Report*, ND-GAIN, 2015.

[69] *Ibidem*.

[70] European Investment Bank, *2019-2020 EIB climate survey*, accessed on the 6th of May 2021.

in any way related to the increasing complexity of global challenges due to greater interconnectedness and globalisation is for specialists to say. However, the success of populist parties has been linked precisely to their proposed simple solutions to complex problems. In this sense, the increasing complexity of the world therefore seems to disorientate modern man.

The comparison with Migration Period, therefore, should not be seen as a direful projection of what will happen to Europe: on the other hand, the analysis aims to suggest what could happen if we do not act on climate change: rising tensions, increase in migratory flows and more frequent pandemics. Moreover, the analysis showed that people from different historical periods behave and react to pandemics and climate change in similar ways, hence the importance of doing such a comparativist work with the aim to provide a model that can explain the interconnections between climate change, pandemics and migrations and society's response to these challenges.

3.2 Are we living in another Migration Period? Possible solutions

Since we have seen that the current anthropogenic climate change might be related to pandemics and migrations, we must now concentrate on the strategies to deal with it. Remarkably, the main difference between the current climate change and the one of the Migration period is that the first one is anthropogenic, so we can mitigate the effects of global warming. However, as stated by Mastrojeni and Pasini, this does not imply that we can return to a pre-global warming situation: indeed, since the situation is already deteriorated, we must anyway prepare to face the future effects of the damages we are currently doing to the environment. Nonetheless, mitigating actions could reduce the effects of global warming in the long run. First of all, we must emphasize the fact that anthropogenic activities have caused so far an increase of 1.2 °C above the pre-industrial level and this is likely that we will reach 1.5 °C between 2030 and 2052 if we do not cut emission rates[71]. Clearly, as we are dealing with a global issue, global collaboration is required. The abovementioned International Panel on Climate Change was indeed created in 1988 to provide world- wide governments with scientific knowledge, but it was the foundation of the UN Framework Convention on Climate Change in 1992 which provided a new framework for discussing the stabilization of the greenhouse gases concentrations in the atmosphere. The Kyoto protocol, Doha amendment and Paris agreement were negotiated in this context. In particular, the Paris agreement was aimed to limit the temperature increase to 2 °C by 2100 while pursuing efforts to reduce it to 1.5 °C. According to the agreement, these targets must be achieved by enforcing national measures aimed to cut the carbon emissions and also with obligations by developed countries to promote sustainable development and establish adequate mitigation and adaptation support measures in vulnerable countries[72].

---

[71] Fawzy, S., et al., *Strategies for mitigation of climate change: a review*, Environmental Chemistry Letters, 18, 2069-2094, 2020.

[72] *Ibidem*.

However, the path towards a global reduction of carbon emissions is not easy. Nowadays, the States with more emissions are economic giants such as China (28% of the total), USA (15%), India (7%), Russia (5%), Japan (3%), and Germany (2%). The UK, France, Italy and Poland each account for 1% of the total account[73]. Moreover, the issue of emission reductions divides the poor from the rich: in fact, the top 1% of the world population by income accounts for about 15% of global emissions, the richest 10% is responsible of nearly the half of them while the poorest 5% only accounts for 6% of them[74]. We should also reflect on the fact that the debate on climate change is not unrelated to geopolitical considerations: Brazil, South Africa, India and China believe, indeed, in the "combined but differentiated responsibilities", whereby states at different stages of development have different responsibilities in the global effort[75], implying that Western States must play the major part in global reduction. However, the 2017 decision of the USA to leave the Paris Agreement has demonstrated further that a shared solution to greenhouse gases emissions seems politically impossible in such an international situation[76].

In this context, the current pandemic of Covid-19 has underlined the urgency and the importance of global cooperation in the face of transnational issues such as pandemics and climate changes. However, a 2021 study claims that there was not a real international cooperation during the pandemic of Covid-19, the response of the states being predominantly motivated by the national interest[77]. Moreover, the WHO has been widely criticised for its role during the sanitary emergency, but we must also bear in mind that like other international organizations, the WHO is at the centre of geopolitical disputes; the global pandemic response, indeed, further emphasises that "geopolitical framework is essential to understanding the capacity and willingness of states and the public to engage with super-wicked problems" such as climate change[78]. Intriguingly, Professor of Physics Antonello Pasini and Professor of Biology Fulvio Mazzocchi argue that there is a similarity between the evolution of the Covid-19 pandemic and the one of climate change: the little and limited number of cases made many argue that the disease could have been brought under control; at the same time, we are only seeing some limited effects of the climate change, due to the inertia of the global climatic system[79]. Thus, national authorities must learn a lesson from the subsequent evolution of the pandemic and provide effective policies to contrast global warming.

---

[73] Union of concerned scientists, *Each Country's Share of CO2 Emissions*, accessed on 21st of May 2021.

[74] Wagstyl, S., *Climate change is becoming less a battle of nations than rich vs poor*, Financial Times, 21st of May 2021.

[75] Dalby, S., *Climate Change and Geopolitics*, Climate Science, 2017.

[76] *Ibidem*.

[77] Basrur, R., Kliem, F., *Covid-19 and international cooperation: IR paradigms at odds*, SN Social Sciences, 1, 7, 2021.

[78] Cole, J., Dodds, K., *Unhealthy geopolitics: can the response to COVID-19 reform climate change policy?*, Bull World Health Organ., 99, 2, 148-154, 2021.

[79] Pasini, A., Mazzocchi, F., *Perception and risk of Covid-19 and climate change: Investigating analogies in a common framework*, Global Sustainability, 3, E32, 2020.

According to the IPCC, in order to meet the 1.5 °C target by 2100, a 45% drop in global emissions from 2010 levels must be achieved by 2030, while for the 2 °C target, states would have to cut world emission by 25% by 2030, in addition to reaching zero-net emissions by 2070[80]. In a recently published report the IPCC stated that if the current trend of the emissions will continue, there will be only a 1% reduction from 2010 levels by 2030[81]. However, the main outcome of the recent COP26 in Glasgow in this respect was that countries declared that they "will pursue efforts to limit the temperature increase to 1.5 degrees C", but whether this will happen can only be seen from next year[82].

Mitigation strategies: a short review

According to a 2020 study published by the Environmental Chemistry Letters, there are three main mitigation strategies to cope with climate change. First of all, as human energy-related activities are the main cause of the increased greenhouse gases concentration, the main strategy is to focus on decarbonization through the introduction of renewable energies and renewable fuels. As of 2018, renewable energy represented the 26.2% of the global electricity production[83].

Furthermore, negative emissions technologies such as afforestation and reforestation, and bioenergy carbon capture are also adopted to contrast the rising CO2 emissions; ocean fertilization, ocean alkalinity enhancement, wetland reconstruction and biochar are other possible strategies[84].

Reforestation strategy is based on the fact that trees capture CO2 while growing, thus being a biogenic negative emission technology. Bioenergy carbon capture, on the other hand, captures CO2 from the atmosphere and store them in geological reservoirs; however, this technology also requires large amount of biomass feedstock and can therefore have a negative impact on the environment. Biochar too has both positive and negative effects and does the carbon capture and storage (CCS), which has significant costs and is highly energy intensive. This technology, indeed, is based on the use of chemicals in order to remove atmospheric CO2 directly from the air. In addition, ocean fertilization is the process of promoting biological activity in the Oceans based of the absorption of CO2 by plankton. However, side effects include unpredictable impacts on food cycles, various effects on the sea ecosystems and even possible increase in the production of greenhouse gases. Wetland restoration, on the other hand, does not present side effects, though a limited number of wetlands have been restored so far[85].

Therefore, given the difficulties that are related to negative emissions technologies, the best approach so far seems to reduce carbon emissions and reforest lands.

---

[80] Fawzi, S., et al., *Strategies for mitigation of climate change: a review*, cit.

[81] UN Climate change, *"Climate Commitments Not on Track to Meet Paris Agreement Goals" as NDC Synthesis Report is Published*, 26th of February 2021.

[82] Mountford, H. et al., *COP26: Key Outcomes From the UN Climate Talks in Glasgow*, WRI, 17th of November 2021.

[83] Fawzy, S, et al., *Strategies for mitigation of climate change: a review*, cit.

[84] *Ibidem*.

[85] *Ibidem*.

3.2.1 A possible long-term solution: The European Green Deal

In a divided global context such the one we described above, it seems unlikely that sovereign states will bind themselves to cut their emission. On the other hand, the European Union aims to be climate-neutral by 2050: this is the core of the "European Green Deal", an action plan presented in December 2019 which seeks to cut pollution, restore biodiversity, and move to a circular and clean economy. These targets are to be achieved in several ways; the EU plans to invest in environmentally friendly technologies, decarbonise the energy sector, provide cleaner forms of private and public transport, ensure more energy-efficient buildings and support industry to innovate in a "greener" way. In order to achieve the Green Deal's targets, the Commission aims to work with the Members states to ensure that the policies linked to the Green Deal's implementation are effectively fulfilled. It should also be underlined that between 1990 and 2018 the EU has reduced greenhouse gases by 23%, while its industry grew by 60%, thus demonstrating that a green economy is possible[86].

Moreover, the "green transition" wishes to be an opportunity to increase the market of sustainable and job-intensive activities; Europe, indeed, recognises that energy-intensive industries (steel, chemicals and cement) are necessary for the economy, but also claims that their decarbonisation and modernisation is vital. A circular economy plan will prioritise reducing and reusing materials and will set minimum requirements to prevent products to be harmful to the environment. Furthermore, by 2030, all packaging must be reusable and recyclable. Business will be encouraged to offer durable, reusable and repairable products; "green" companies will also have to comply with several standards[87].

Internal transport, as said, must be sustainable, and this will also be true for inland freight, which will be more by rail and inland waterway rather than by road. Moreover, one million recharging stations will be available by 2025 for the 13 million zero and low emissions vehicles; furthermore, there will be also more stringent air pollutant emissions standards[88].

A biodiversity strategy will also be aimed to increase the protected areas, which play an important role in mitigating natural disasters; the European Commission will also pursue reforestation and afforestation in order to offset CO2 emissions. Resilient seas and healthy oceans will also be a priority of the ecosystem strategy; moreover, Europeans authorities have stated that they will have a zero toler ance for illegal and unreported fishing[89].

In order to achieve this ambitious plan, a 260 billion annual investment will be needed; a Just Transition Mechanism will also be dedicated to those regions which depend the most on fossil fuels and carbon intensive activities. The aim is to achieve a socially just transition and leave no region behind[90].

---

[86] *Ibidem*.

[87] European Commission, Communication from the Commission to the European Parliament, the European Council, The Council, the European Economic and Social Committee and the Committee of the Regions - *The European Green Deal*, 11th of December 2019.

[88] *Ibidem*.

[89] *Ibidem*.

[90] *Ibidem*.

In conclusion, the EU is aware of the risk of carbon leakage, which can happen in two ways: production can be shifted to countries with less stringent carbon emissions controls or by replacing EU products with more carbon-intensive imports. This would frustrate EU efforts: the Commission therefore plans to propose a "carbon border adjustment mechanism"[91].

These efforts to make Europe "the first zero emission Continent", however, might also be frustrated by MSs. Recently, some critical voices have argued that the Emission Trading Scheme (the key tool of the EU's strategy for reducing the greenhouse gases[92]) should be discontinued, while other have insisted on abandoning or cancelling[93] the Green Deal[94]. Moreover, according to critics, the Green deal does not call on citizens to change their attitudes, which makes it unlikely that they will start changing their behaviours on their own; in addition, the text of the Green Deal has been accused of being ambiguous in the sense it does not clearly identifies the very agents of the proposed actions.

On the other hand, several researchers argue that the European Green Deal is a "great opportunity", though they agree that in order to make it successful, "it must be anchored in the concepts pertaining to the constitutional framework of the EU legal order [...] the concepts of solidarity, sustainable development and high level of environmental protection"[95].

Moreover, other authors link the European Green Deal to the issue of social and economic inequalities: according to a study published in the Journal of Eastern Europe Research in Business and Economics, the EU have the potential to become the world leader in the green transition if it manages to address social and economic inequalities along with the climate change issue[96]. New green sectors, indeed, can provide employment opportunities for unemployed young people, as also stated by another study published on The Lancet[97].

In conclusion, the European Green Deal can potentially make Europe a "greener" continent, but differences among MSs must be overcome; problematic issues such as the refugee crisis and the pandemic of Covid-19 have already posed problems to states and a delicate issue such as "green transitions" can possibly divide Europe. In this sense, we must recall the recent debate on a "multi-speed Europe": according to some studies, this solution could allow those MSs that are willing to continue European integration to increase the efficiency of the European policies, while allowing other states to pursue their own national policies[98]. Conversely,

---

[91] *Ibidem*.

[92] *Ibidem*.

[93] European Commission, *EU Emissions Trading System (EU ETS)*, accessed on 22nd of May 2021.

[94] Elkerbout, M., *The European Green Deal after Corona: Implications for EU climate policy*, CEPS, 6, 2020.

[95] Eckert, E., *Sustainability in the European Union: Analyzing the Discourse of the European Green Deal*, Journal of Risk and Financial Management, MDPI, 2021.

[96] Sikora, A., *European Green Deal - legal and financial challenges of the climate change*, ERA Forum, 21, 681-697, 2021.

[97] Nae, T., Panie, N., *European Green Deal: The Recovery Strategy Addressing Inequalities*, Journal of Eastern Europe Research in Business and Economics, 2021.

[98] Becchetti, L., et al., *European Green Deal as social vaccine to overcome COVID-19 health & eco-*

other argues that this solution would be an error, breaking Europe's strength: its unity[99]. However, Europe risks finding itself in the same situation as the international arena: it risks being torn between different national interests and opting for a compromise that would not help to resolve such important issues. On the contrary, when dealing with such critical issues, rapid responses are needed. Furthermore, European states risk being among the few states that will unilaterally decide to curb the emissions; that is why a "green diplomacy" and an international climate finance[100]are needed, as also emphasized by the European Commission[101]. In this sense, a stronger and more united Europe would serve as an example to the international community.

If the European Green Deal actually reduce the European emissions, the results will be evident in the long-term period. However, we are already facing huge migratory fluxes. What should be Europe's approach for the short and medium term?

3.2.2 Short and middle-term policies

Broadening the refugee status

According to the 1951's Refugees Convention, a refugee is someone who "is unable or unwilling to return to its country of origin owing to a well-founded fear of being persecuted for reasons of race, religion, nationality, membership of a particular social group, or political opinion"[102]. Recently, however, a growing number of publications stressed the need to include in the refugee definition also "people leaving their countries due to extreme climatic event"[103]. Those who are not defined as refugees such as "economic migrants" indeed, do not possess the credentials to benefit from the international protection granted by the 1951 Convention and their position must be determined by the host state[104]. Already in 1990, the IPCC stated that climate change would lead to more people migrating, but to date the definition of refugee has not been changed; however, according to Behrman and Kent, University of London and University of East Anglia, the same legal framework that applies to political refugees (the responsibility of the international community when people are no longer protected by their states – must be used for the climate refugees, as the state from which they come is no longer able to protect them as it lacks resources, is hit by extreme weather events or is disappearing (as in the case of the islands)[105]. As the ef-

nomic crisis.

[99] Kundera, J., T*he Future of EU: Towards a two Speed Europe*, European Research Studies Journal, 22, 3, 261-281, 2019.

[100] Sobotka, B., T*wo-speed Europe is a mistake*, Politico, 6th of December 2017.

[101] European Commission, *International climate finance*, accessed on 23th of May 2021.

[102] European Commission, *EU climate action and the European Green Deal*, accessed on 23th of May 2021.

[103] UNHCR, *Text of the 1951 Convention Relating to the Status of Refugees*, 1951.

[104] Behrman, S., Kent, A., *Climate Refugees: Beyond the Legal Impasse?* Taylor and Francis, ebook, 2018.

[105] UNHCR, *UNHCR Resettlement Handboo and country chapters, Refugee status and resettlement*, 2018.

fects of climate change are likely to worsen, we can argue that more and more people will find themselves in need of international protection: establishing the status of climate refugee could therefore be an incentive to address the issue of climate change in international *fora*.

Avoiding deaths in the Mediterranean: the humanitarian corridors

In addition to granting the international protection to climate refugees, however, other policies must be pursued. In the short-term, indeed, European authorities must address the issue of refugees that will arrive, or which have already arrived. In order to avoid humanitarian disasters such as the shipwrecks we hear about every day, alternative solutions to dangerous Mediterranean crossing the must be provided for refugees who have the credentials to obtain the international protection. In the current situation, indeed, even those who are eligible for the status of refugee have no choice but to risk the dangerous crossing of the Mediterranean so to be recognized as refugees in Europe; in order to respond to such a situation, in recent years some private organizations have begun to work on a solution, the *humanitarian corridors*, which provides legal entry to Europe for those who possess the refugee credentials[106]. The humanitarian corridors indeed, are based on article 25 of the European Visa Code, which allows states to grant visas for humanitarian or national interest reasons; refugees apply in so called *transit countries* (Jordan, Lebanon, Ethiopia) and are brought to Europe via plane. Then, once arrived, they are immediately welcomed in local communities to avoid marginalization, while they are also initiated to employment[107]. This approach is also considered a disincentive for the dangerous crossing but it also help counteract illegal human trafficking[108]. So far, humanitarian corridors have brought people to Italy, France, Germany, San Marino, and Andorra from Ethiopia, Jordan, Lebanon and Greece, but also from Mexico to the USA[109]. In the face of increasing migration flows, humanitarian corridors could be a viable alternative to agreements with migrants' countries of departure.

Integrating migrants into the labour market

The humanitarian corridors approach reminds us of the resettlements of the Germanic tribes into the Roman Empire; in fact, based on what we have seen, we can understand that these populations' transfers were motivated by the absence of sufficient labour. Historians, indeed, argue that these people were settled in "deserted lands"[110]. Interestingly, we have also previously stated that Europe is now facing a

[106] Behrman, S. and Kent, A., *Climate Refugees: Beyond the Legal Impasse?* cit.

[107] Ricci, C., *The Necessity for Alternative Legal Pathways: The Best Practice of Humanitarian Corridors Opened by Private Sponsors in Italy*, German Law Journal, Cambridge University Press, 21, 265-283, 2020; Rolando, F., Naso, P, *Humanitarian Corridors to Italy: An Interview With Professor Paolo Naso*, Harvard International Review, 39,2, 64-67, 2018.

[108] Rolando, F., Naso, P, Humanitarian Corridors to Italy: An Interview with Professor Paolo Naso, cit.

[109] *Ibidem*.

[110] Sant'Egidio, *Humanitarian corridors*, accessed on 24th of May 2021.

demographic decline; the majority of EU's MS, indeed, will experience declining labour force size, a process which will accelerate in the following decades[111]. By 2050, the European working age population (20-54) will decrease by 49 million, while in the USA it will increase by 20 million, and most importantly, it will reach 1.3 billion in Sub-Saharan Africa[112]. In this sense, a study published by the European Union argued that a more labour integration of immigrants would decrease the expected decline in the employed population by half, while, on the other hand, non-integration might double the decline[113]. Moreover, according to another study, in Southern European countries in particular, immigrants are already highly employed in the shadow economy, which, in 2002, accounted for "28.3 per cent of GDP in Greece, 26.2 per cent of GDP in Italy and 22.3 per cent of GDP in Portugal and Spain"[114]. According to the authors of the cited study, the number of undocumented migrant workers would be in the hundreds of thousands in these countries. Moreover, according to a recent study, extra-UE immigrants already make up 13% of key workers (teaching professionals, skilled agricultural workers, science and engineering associate professionals, personal care workers and cleaners and helpers), but they are also overrepresented in many European countries in this category, especially among low skilled key workers[115].

It is becoming evident, indeed, that Europe needs immigrants, as was already stated in a 1992 paper: the inflow of immigrants would ensure population stabilization, but would also increase the working age population[116]. Interestingly, a 2019 proposal put forward by a study published by the Centre for European Policy Studies called for the creation of an "EU-Africa partnership for human capital formation and skill mobility"[117]. The authors argued that Europe should fund vocational training programmes in Africa based on African and European labour market needs, issue working permits and involve the private sector to facilitate job seeking across the Mediterranean[118]. Such a policy of integrating migrant workers would help boost the European economy, while giving migrants the opportunity to develop labour skills that could also support a potential future business activity in their

---

[111] Mathisen, R., *Peregrini, Barbari, and Cives Romani: Concepts of Citizenship and the Legal Identity of Barbarians in the Later Roman Empire*, cit.

[112] Marois, G., Potancˇoková, M., *Scenarios of labour force participation and employment integration of immigrants in the EU: demographic perspective*, JRC Technical report, European Commission, 2020.

[113] European Commission, *Demographic trends of workforce*, accessed on 24th of May 2021

[114] Marois, G., Potančoková, M., *Scenarios of labour force participation and employment integration of immigrants in the EU: demographic perspective*, cit.

[115] Peixoto, J., et al., *Immigrants, markets and policies in Southern Europe*, in Immigrants, markets and policies in Southern Europe, Trends, Structures and Policy Implications, edited by Marek Okólski, IMISCOE Research, Amsterdam University Press, 2012.

[116] European Commission, *Immigrant Key Workers: Their Contribution to Europe's COVID-19 Response*, 24th of April 2021.

[117] Coleman, D., *Does Europe Need Immigrants? Population and Work Force Projections*, The International Migration Review,26, 2, Special Issue: The New Europe and International Migration,413-461, 1992.

[118] Barslund, M., et al., *EU-Africa partnership for human capital formation and skill mobility*, CEPS, MEDAM, 2019.

countries[119]. In this sense, this approach is linked to possible European investments in Third World countries, in order to create the conditions for a sustainable economic growth that can prevent people from migrating.

Boosting the Third World (green) economy

Giving refugees the chance to enter Europe legally and safety is a short-term solution; on the other hand, migrants and refugees should also be given the possibility to stay in their countries and to find work and study opportunities there. As stated in the previous paragraph, in fact, Europe is already the world's largest contributor of development aid and the largest donor of financial aid to mitigate climate change, as 20% of its current budget is dedicated to climate finance[120]. Investing in rural areas would help people not to look for job opportunities elsewhere: in addition, Africa contains half of the world's uncultivated land that could be used to produce food, but spends 35 billion on food imports every year[121]. However, Africa's agricultural export potential is enormous, as the continent could export – for example – to the neighbouring Middle East region, which is predominantly dry land[122]. On the other hand, it is also important to create the conditions for the integration of developing countries into the world market: FDI is essential to stimulate economic growth and exports[123]. However, it is also important to remember, as mentioned above, that trade can help reduce poverty, but economic underdevelopment in these countries is also fostered by widespread corruption, bad governance and insufficient protection of human rights, among other factors. Policymakers should therefore also consider this aspect.

In addition, developed countries must also help developing countries to create the basis for sustainable development, from a social, economic and environmental perspective. This is, for example, the approach of the cooperation between the European Union and the African Union, which aims to improve market access for young farmers and smallholders, with the objective of supporting the development of agribusiness. Moreover, as recently stated by EU Commission President Ursula Von der Leyen, "the potential for the green economy in Africa is enormous" and green transition can create jobs and stimulate sustainable growth in the post-pandemic era, both in Africa and Europe[124]. Furthermore, according to UNCTAD, green energy would help developing countries to be less dependent on the volatility of the international market[125]. Furthermore, a comprehensive study, "Inclusive Green Growth.

---

[119] *Ibidem.*

[120] UNCTAD, *Promoting refugee and migrant entrepreneurship in East Africa, the Andean region, and the Middle East*, accessed on 25th of May 2021; Fondazione ISMU, *Migranti africani fanno impresa grazie a un percorso di formazione e mentoring – Comunicato stampa* 12.10.2020, 12th of October 2020.

[121] European Commission, *International development aid*, accessed on the 10th of May 2021.

[122] IFAD, *Investment in Africa's rural areas will curb economic migration, IFAD President tells ministers*, 04th of July 2016.

[123] LCLUC, Dryland Degradation in the Middle East, accessed on 25th of May 2021.

[124] European Commission, Communication from The Commission to The European Parliament, The Council and the European Economic and Social Committee, Trade, growth and development, 27th January 2012.

[125] European Commission, *EU-Africa Business Forum: over 3 000 participants focus on the transition*

Advances in African Economic, Social and Political Development", concluded that green business in Africa improves the quality of life in rural areas of the continent, stimulates innovation, provides job creation and poverty reduction[126]. This is also confirmed by the report published by UNEP, "*Green economy, developing countries success stories*", according to which the green economy and sustainable urban planning have beneficial effects in terms of economic growth in developing countries[127]. However, it is also important that these policies take in account local climate change adaptation practices[128], in order to be more successful.

The importance of sustainable growth lies in the fact that it creates job opportunities in the intermediate period and diminishes the effects of climate change in the long run. However, mitigation strategies must always go hand in hand with development projects: in fact, as early as 1988, a study published by the US National Academy of Science argued that protected areas will not survive in a human environment that struggles with hunger[129]. In fact, forests are burned to produce food for families, so it is necessary both to provide alternative forms of livelihood and to save biological diversity. For example, the same study suggests the use of multiple tree species in agricultural activities, mixing trees with food crops and increasing environmental education[130], thus providing an example of the kind of approach needed.

Urbanization planning in developing countries

Massive urbanisation, as we have seen, is considered to be one of the main causes of biodiversity loss and, hence, climate change; however, this is not only true for developed countries, but increasingly for developing ones[131]. Importantly, it is predicted that by 2030 there will be almost 2 billion more urban dwellers, most of them in small towns in developing countries: this urbanisation will therefore have great effects on humanity and the environment[132].

Developing countries have fewer resources to devote to land protection in cities and therefore biodiversity loss may not be a priority[133]; however, as these countries are expected to host most of the new population in the years to come, the interna-

to green and sustainable energy in Africa, 28th of April 2021.

[126] UNCTAD, *Green energy could help developing countries diversify their economies*, 12th of October 2017.

[127] Ngondjeb, D., et al., *Insights on Africa's Future in its Transition to the Green Economy*, in: Atewamba C., Yong Ngondjeb D. (eds) Inclusive Green Growth. Advances in African Economic, Social and Political Development. Springer, Cham, 2020.

[128] Sukhdev, P., et al., *Green economy, developing countries success stories*, UNEP, 2010.

[129] Nyong, A., et al., *The value of indigenous knowledge in climate change mitigation and adaptation strategies in the African Sahel*, Mitigation and Adaptation Strategies for Global Change, 12, 787-797, 2007.

[130] Brady, N., *International Development and the Protection of Biological Diversity, in Biodiversity*, Wilson E.O. editor, National Academies Press, 1988.

[131] *Ibidem*.

[132] Pauchard, A., *Multiple effects of urbanization on the biodiversity of developing countries: The case of a fast-growing metropolitan area (Concepción, Chile)*, Biological Conservation, 127, 3, 272-281, 2006.

[133] McDonald, R., et al., *The implications of current and future urbanization for global protected areas and biodiversity conservation*, Biological Conservation, 141, 6, 1695-1703, 2008.

tional community needs to consider possible policies to prevent huge destruction of the natural habitat. *"The Future of Global Urbanization and the Environment"* by McDonald et al, argues that possible policies could include treating ecosystems as urban services, making a global effort to protect biodiversity during the urbanization process, and coordinating internationally to provide guidelines for urban sustainability[134].

Increasing urbanization is not only linked to the possible disruption of habitats, but also to bushmeat consumption. Which, we have seen, is linked to zoonotic spill over. Importantly, according to McDonald et al., the bushmeat trade can possibly increase in West Africa due to the proximity of natural habitats to cities that will grow in the near future[135]; moreover, another study claims that trade of parts of animals has recently increased and the high demand of the global market, as well as the current patter of urbanization will influence the habitat loss and degradation[136]. Population growth will also increase the demand for bushmeat; however, importantly, bushmeat consumption has also been linked to poverty and political instability (as lack of law enforcement reduces the availability of alternative food)[137].

In order to act on the bushmeat hunting, indeed, it is necessary to distance people from wildlife: protected areas must be promoted and human settlement within there must be limited, while human movement within the protected areas must be controlled. Moreover, land zoning has been successful in reducing bushmeat hunting in areas where human settlements bordered reserves[138].

3.2.3 Preventing the next pandemic

As said, preventing habitat disruption aims to prevent bushmeat consumption which, in turn, aims to prevent possible zoonotic spill over. Policies must therefore address the causes of wildlife consumption, or habitat disruption, which, as we have seen, also include poverty and lack of food. However, this approach must also be accompanied by the abolition or the limitation of wildlife trade, as China has recently done in the wake of the Covid-19 pandemic[139]. According to Kolby, J., National Geographic, "clean trade" could reduce the risk of spreading pathogens: this would include testing before transport and at the border, as it usually done for cattle[140]. Surveillance of wildlife for high-risk pathogens and surveillance and risk reduction in

---

[134] Elmqvist, T., et al., *Urbanization, habitat loss and biodiversity decline, solution pathways to break the cycle*, in: K.C. Seto, W.D. Solecki, C.A. Griffith (Eds.), The Routledge Handbook of Urbanization and Global Environmental Change. Routledge, Abingdon, Oxon, UK, 139-151, 2016.

[135] McDonald, R., et al., *The Future of Global Urbanization and the Environment*, Solutions, 2016.

[136] McDonald, R., et al., *The implications of current and future urbanization for global protected areas and biodiversity conservation*, cit.

[137] Güneralp, B., *Urbanization in Africa: challenges and opportunities for conservation*, Environmental Research Letters, 13, 1, 2017.

[138] Lindsey, P., *The bushmeat trade in African savannas: Impacts, drivers, and possible solutions*, Biological Conservation, 160, 80-96, 2013.

[139] *Ibidem*.

[140] Mallapaty, S., *China set to clamp down permanently on wildlife trade in wake of coronavirus*, Nature, 21st of February 2020.

people at high risk of contact with wildlife must also be enforced[141]. A global effort is also needed: in the case of Covid-19 the virus came from China, but other "emerging disease hotspots" are also South Asia, sub-Saharan Africa, and Latin America: international sanitary organization should therefore improve surveillance in these areas[142]. An holistic approach which combines surveillance and control, reinforced health system, ecosystem management and collaboration with social sciences, could be a possible solution to prevent the next pandemic[143].

Moreover, epidemiologist Michael Osterholm and author Mark Olshaker, argue that SARS, MERS, the "swine flu", Ebola and Zika should have been a "wake up call" for international authorities, but they were not. However, a rapid response could help limit the spread of a potential virus. In 1918 cities which reacted early to the flu pandemic had less casualties; however, this requires reliable information and responsiveness from the outset, which was lacking in several cases. Interestingly, in October 2019, The Johns Hopkins Centre for Health Security hosted a high-level pandemic exercise – Event 201 – which highlighted the need for cooperation among industries, national governments, and International Organizations[144]; however, some commentators argue that this event and past ones demonstrated a general lack of coordination and confusion[145]. Moreover, the article claims that Covid-19 should not be considered a "one in a hundred year event"[146], for the reasons we have already analysed: an increasingly connected world, rapid urbanization creating conditions for malnutrition and poor hygiene, and millions of people living in contact with animals.

Furthermore, we must also keep the medical aspects in mind: as we saw during the early stages of the pandemic, states were competing with each other to secure the necessary medical devices; domestic production of vaccines and health technologies should be therefore promoted globally to respond to the immediate needs of the population. For instance, Europe has recently announced that it will promote the manufacturing and access to vaccines, medicines and health technologies in Africa[147]. Similarly, it is also necessary to support health education on each continent and adequate logistics to enable all types of medicines to reach different parts of the globe.

Ultimately, international cooperation is needed: Osterholm and Olshaker, indeed, argue that smallpox was eradicated as a result of the US and the USSR's decision to collaborate[148]. Cooperation is indeed necessary, especially in the health sec-

---

[141] Kolby, J., *To prevent the next pandemic, it's the legal wildlife trade we should worry about*, National Geographic, 7th of May 2020.

[142] Daszak, P., et al., *A strategy to prevent future epidemics similar to the 2019-nCoV outbreak*, Biosafety and Health, 2,1, 6-8, 2020.

[143] *Ibidem*; Roche, B., et al., *Was the COVID-19 pandemic avoidable? A call for a "solution-oriented" approach in pathogen evolutionary ecology to prevent future outbreaks*, Ecology Letters, 23, 11, 1557-1560, 2020.

[144] Roche, B., et al., *Was the COVID-19 pandemic avoidable? A call for a "solution-oriented" approach in pathogen evolutionary ecology to prevent future outbreaks*, cit.

[145] Center For Health Security, Event 201, accessed on the 26th of May 2021.

[146] Osterholm, M., Olshaker, M., *Chronicle of a Pandemic Foretold*, cit.

[147] *Ibidem*.

[148] European Commission, *€1 billion Team Europe initiative on manufacturing and access to vaccines, medicines and health technologies in Africa*, 21st of May 2021.

tor and in the exchange of information: international conferences aimed at studying the new annual strains of influenza and planning the production of vaccines are already in place, but this kind of cooperation must be strengthen. It must be stressed that, although there are more IOs than during the Spanish flu pandemic, states have still adopted unilateral approaches; however, this does not mean that international cooperation is doomed to fail, but rather that there is a reluctance or ambivalence about cooperating at international level[149]. This is exemplified, for example, by the WHO itself, to which states have assigned increasing tasks while limiting its independence. According to Stewart Patrick, Senior Fellow in Global Governance at the Council of Foreign relations, "multilateral institutions are what states and their leaders make of them": this is, at least, the lesson drawn from the G-20 response to the 2008 economic crisis, which was coordinated by President Obama and helped to avoid another Great Depression[150]. Such a lack of cooperation would have not happened in a cosmopolitan world, Patrick argues, but in recent years democracies have faltered under the blows of nationalism and populism, thus neglecting international cooperation[151]. Furthermore, a reluctance to cooperate internationally has gone hand in hand with the unwillingness to implement commitments such as those requested by the WHO, which, in turn, has little influence on national governments.

In conclusion, we agree with the author of "*When the system fails*": governments are interested in a multilateral system which can provide an effective response to crises such as Covid-19 but are less willing to give up to their sovereignty in a issue as delicate as health[152]. However, this analysis has shown that international cooperation is necessary in such an interconnected world, whether it is climate change, refugees, the economy or pandemics. Until states are willing to cooperate fully on these issues, we are likely to continue to experience the effects of climate change, wars, migrations, and pandemics.

3.3 Avoiding another Migration period

In the previous paragraph we looked at possible policies that could help prevent thousands of deaths in the Mediterranean, prevent people from leaving their homes, stimulate "green" economic growth, avoid another pandemic and mitigate the effects of climate change. Studies have shown that we can effectively mitigate climate change. We can therefore limit its disruptive effects: it's a matter of cooperation. Above all, it is important to have analysed existing policies and best practices which need to be strengthen and promoted, along with climate mitigation actions: these policies and actions have a great potential, but require a global effort to have sustainable and lasting results. The European Green deal has yet to be approved and is already facing oppositions; humanitarian corridors have only been implemented by a few European countries and by Canada and the US only; green and migrant start-

149 Osterholm, M., Olshaker, M., *Chronicle of a Pandemic Foretold*, cit.

150 Patrick, S., *When the system fails, Foreign Affairs*, 99, 4, 40-51, 2020.

151 *Ibidem*.

152 *Ibidem*.

ups are still small and few in number so far, and the current pandemic of Covid-19 has highlighted the general reluctance to cooperate internationally. Nonetheless, if Europe overcame its divisions and committed itself to reducing emissions, it could set an important example for the international community; moreover, we claim that best practices at the micro-level and policies such as the humanitarian corridors and the migrant business provide important approaches which can be easily followed. We have argued before that the current global interconnection facilitates, for example, the spread of pandemics, but it can also facilitate the adoption of these best practices. Indeed, social networks, the very symbols of our global era, are becoming increasingly important for conveying messages and examples: this was evident in the aforementioned *Arab springs*, but also in the global youth movement "Fridays for future". Such global movements have the potential to change and influence the global agenda and will certainly play an important role in the near future.

This last chapter, therefore, aimed not only to provide possible policies, but also to show that we already have policies that can meet epochal changes. In presenting the entire study, we claimed that climate change might provoke pandemics, conflicts and migrations. However, studies also claimed that even though conflicts (and thus migrations) are caused by a variety of situations, they are more likely to take place in countries with low human development[153] – which can easily enter in a conflict trap – or in countries affected by horizontal inequality[154]. Therefore, the above policies could not only prevent migration, but also conflicts, thus creating a positive spiral between economic growth and climate change mitigation. Moreover, trade has been also associated to peace in the sense it not only creates economic growth and interdependence, but according to political philosophers, it also favours the understanding of each other's culture[155]. Trade, therefore, might also increase international cooperation, while at the same time increasing global economic growth; international cooperation, on the other hand, is what is necessary to address global issues such as the addressed earlier. On the other hand, according to Nicholas Stern, former World Bank Chief Economist, climate change could also provoke side effects on society and international cooperation, thus reducing global GDP by 20%[156]. It is therefore in everyone's interest to avoid drastic climate change, even if only from an economic point of view.

In conclusion, we may argue that we already got the keys to avoid another Migration Period. To sum up, we could say that both international cooperation and best practices are needed: in such a globalized world, indeed, both a small virus and local best practice can have a worldwide impact.

Conclusions

The current study was guided by a consideration which humanists usually trace to a phrase from Cicero's De Oratore: "Historia vero testis temporum, lux veritatis,

153 *Ibidem*.

154 Humphreys, M., *Economics and violent conflicts*, Harvard University, 2003.

155 *Ibidem*.

156 *Ibidem*.

magistra vitae, nuntia vetustatis": History [is] the witness of time, the light of truth,the life of memory, the directress of life, the herald of antiquity[157]. In other words, history is often worth studying because it could help to grasp modern trends, predict possible evolutions and illuminate processes through the lens of past events. The research question "Are we living another migration period?" was therefore inspired by the desire to approach the study of a period such as the contemporary one that can be defined as one of crisis or, at least, of epochal change. As was the case with the migration period.

The choice to focus on the migration period was motivated mainly by two reasons. First of all, the period of migrations is also known in Italian and French as the period of the "barbarian invasions". Such a marked difference (invasions/migrations) is particularly interesting because it draws attention to the different interpretation and meaning that men could give to the same historical process. In particular, the "Roman" world clearly saw this period as negative since it meant the end of the Roman Empire, while the "Germanic" world considered the same period as a period of migrations motivated by different factors, as migrations usually are. Given these two opposing interpretations, a natural question arises: how will our period be seen?

Another reason to focus on the migration period was given by new interpretations according to which the movements of the Germanic tribes were also influenced by a change in climate. This new approach puts the focus back on our period: the current climate is also changing, and this will certainly push more people to leave their countries.

A single study could therefore have been devoted only to a comparison between past and present migrations: however, it is interesting to note that both the migration period and the present era are characterised by several pandemics, the repeated outbreaks of the Justinian plague and the various zoonotic diseases of the twentieth century. How then do these diseases interact with migratory movements?

The current study therefore aimed to investigate these links through comparison, in order to predict possible developments and provide policies on how to address these historical challenges.

From the outset, the study aimed to include different perspectives and disciplines in the analysis in order to pursue a holistic approach. Indeed, such an approach is necessary in a globalised world such as the one in which we live. However, even if we cannot properly define the world of the migration period as "globalised" in the current meaning of the term, it actually was, and the current study proves it. Trade routes and social networks are certainly more developed these days than they were in the early Middle Ages, but even then there were connections and social networks. The importance of studying the past from this point of view, in fact, lies in the fact that the world has become increasingly interconnected: the more social networks there are, the more the trends are accelerated. We can argue that the technologies are different between the two periods, but the historical trends do not differ: migrations are still driven by the same factors (violence, economic reasons, attractiveness of the society/country of arrival, changing climatic conditions) and

[157] Mastrojeni, G., Pasini, A., *Effetto guerra, effetto serra*, cit.

can therefore be studied and compared. Pandemics still have the same effects on societies, and we can therefore provide interesting analyses.

As mentioned, therefore, a holistic approach was necessary: in the course of this analysis, we have considered different studies from different branches of science: medicine, epidemiology, sociology, economics, physics, biology and natural history. We argue, in fact, that even if they are sciences of different types (social, humanistic, medical...) they all deal with humanity; in such a globalised world, there is no place for separateness of sciences. Cooperation is therefore necessary. In fact, cooperation is necessary both between the sciences and between nations. Indeed, humanity already possesses solutions to deal with climate change and its effects. However, these solutions have not been fully implemented or are marginal; as we live in the era of online social networks, these best practices have the potential to become global, to be widely discussed and to change the global agenda. The rise of social networks does not only mean a faster spread of viruses, but also a faster and wider spread of information, beliefs and knowledge. Which is what humanity needs in order to meet these historical challenges.

A (more) globalised world, therefore, is both a challenge and a solution. Global interconnections are also at the root of disputes, violence, inequalities; however, investing in the green economy in developing countries, for example, means creating the conditions for a 'green' growth that could both mitigate the effects of climate change and avoid dangerous Mediterranean crossings. In addition, inclusive and sustainable economic growth could also reduce the risk of conflict. Importantly, this does not mean that investing in green business will solve the world's major problems; however, we can argue that it could actually provide the basis for lasting peace. After being the epicentre of two global conflicts within a generation, Europe has not experienced internal wars since the beginning of the process of economic and political integration[158]. Now Europe must commit itself to promoting peace and economic growth in other parts of the world; in fact, this is already happening, but it will only be successful if it also includes cooperation with international actors and if it adopts a holistic approach.

To conclude, we could ask ourselves the question already mentioned: How will our period be regarded? Europe (and the world) is at a crossroads; in the course of this analysis we have seen that the processes that are taking place and that took place in the Migration period are extremely similar, so that they can be studied with a comparative approach. However, since the Migration period the world has become increasingly connected due to increased trade routes, the internet and wider social networks, among other factors. This indeed poses new challenges and solutions: Europe could avoid further pandemics and migration within its borders, but it cannot prevent the effects of climate change from reaching it anyway.

Will Europe then close itself off from the world? A global perspective and a holistic approach, on the other hand, could provide possible solutions.

We are at the beginning of another Migration period: looking back may therefore mean changing our future.

---

[158] Cicero, *On oratory and orators*, Watson, J.S., Harper & Brothers, New York, 1860.

Bibliography

Abikova, J., Piotrowicz, W., *Shaping the Balkan corridor: Development and changes in the migration route 2015-16*, International Migration, IOM, Wiley, 59, 2021

Abshir, S., *Climate change and security in the Horn of Africa: can Europe help to reduce the risk?* European Institute of Peace, 2020

African Union, *Africa labour migration outlook in the post Covid-19 era*, Policy Brief I, 2020

Adugna, G. *Migration patterns and emigrants' transnational activities: comparative findings from two migrant origin areas in Ethiopia*, CMS 7, 5, 2019

Aimone, M., *Romani e Ostrogoti fra integrazione e separazione: il contributo dell'archeologia a un dibattito storiografico,* Reti Medievali Rivista, 13, 1, 31-96, 2012

Affeldt, J., Seaman, M., *A Gateway Re-opens: the growing popularity of the Atlantic route, as told by those who risk it*, MMC Research Report, February 2021

Alden, C., *Emerging Powers and Africa: From Development to Geopolitics,* IAI, 19, 2019

Alföldy, *The Crisis of the Third Century as Seen by Contemporaries,* Greek, Roman, and Byzantine Studies, 15, 89-111, 1974

Ali, F., et al., *Climate Change-Induced Conficts in Pakistan: From National to Individual Level,* Earth Systems and Environment, 2, 573-599, 2018

Anderson, M., *Covid-19 exposes weaknesses in European response to outbreaks*, BMJ, 368, 1075, 2020

Atkinson, J., *The Plague of 542: not the birth of the Clinic,* Acta Classica, Vol. 45, pp. 1-18, 2002

Audoin-Rouzeau, F., Vigne, J., *The black rat (Rattus rattus) in Roman and Medieval Europe: The commercial roads and the spread of plague*, Anthropozoologica, 92, 5, 422-6, 1997

Axboe, M., *The year 536 and the Scandinavian gold hoars,* Medieval Archaeology, Society for Medieval Archaeology, 43, 186-188, 1999

Azhar, E., et al., *Evidence for Camel-to-Human Transmission of MERS Coronavirus*, New England Journal of Medicine, 370, 2499-2505, 2014

Basrur, R., Kliem, F., *Covid-19 and international cooperation: IR paradigms at odds,* SN Social Sciences, 1, 7, 2021

Barbiera, I., Dalla-Zuanna., G., *Population Dynamics in Italy in the Middle Ages: New Insights from Archaeological Findings, Population and development review* 35, 2, 367-389. 2009

Barlow, M., et al., *A Review of Drought in the Middle East and Southwest Asia*, Journal of Climate, 29(23), 8547-8574, 2016

Barret, J., *What caused the Viking Age*, Antiquity, 82, 671-685, 2008

Barslund, M., et al., *EU-Africa partnership for human capital formation and skill mobility,* CEPS, MEDAM, 2019

Bazanova, L.P. et al., *Natural foci of plague in India: epizootological characteristics*, Meditsinskaia Parazitologiia i Parazitarnye Bolezni, 3, 60-63, 2009

Becchetti, L., et al., *European Green Deal as social vaccine to overcome COVID-19 health & economic crisis,* The Lancet, 2, 100032, 2021

Benavides, M., *Evidencias arqueológicas de la peste justinianea en Hispania*, Arqueología y Prehistoria del Interior Peninsular, 8, 55-55, 2019

Berglund, B., *Human impact and climate changes-synchronous events and a causal link,* Quaternary International, 105, 1, 7-12, 2003

Bianchi, G., McCave, I., *Holocene periodicity in North Atlantic climate and deep-ocean flow south of Iceland,* Nature 397, 515-517, 1999

Black, R., et al., *The effect of environmental change on human migration*, Global Environmental Change, 21,1, S3-S11, 2011

Bloom, D., et al., *The cost of low fertility in Europe,* National Bureau of Economic research, 2009

Bosello, F., et al., *Economic impacts of climate change in Europe: sea-level rise,* Climatic Change, 112, 63-81, 2012

Boulnois, L., *La Via della Seta,* Bompiani, Milano, 2005

Bousquet, J., *Is diet partly responsible for differences in COVID-19 death rates between and within countries?,* Clinical and Translational Allergy, 10, 16, 2020

Bozorghmer, K., et al., *COVID and the convergence of three crises in Europe,* The Lancet, 5, 5, 247-248, 2020

Brashares, J., et al., *Wildlife decline and social conflict. Policies aimed at reducing wildlife-related conflict must address the underlying causes*, Science Magazine, 345, 6159, 376-378, 2014

Brooke, J., *The Global Dark and Middle Ages, AD 542-1350,* in Climate Change and the course of Global History: a rough journey, Studies in Environment and History Cambridge University Press, 2014

Brooke, J., *Growth beyond Limits: 1945 to Present*, in Climate Change and the Course of Global History: A Rough Journey, Studies in Environment and History, Cambridge University Press, Cambridge, 2014

Brady, N., *International Development and the Protection of Biological Diversity*, in Biodiversity, Wilson E.O. editor, National Academies Press, Washington, DC, 1988

Brock, W., Xepapadeas, A., *The Economy, Climate Change and Infectious Diseases: Links and Policy Implications, Environmental and Resource Economics*, 76, 811-824, 2020

Broniatowski, D., *Weaponized Health Communication: Twitter Bots and Russian Trolls Amplify the Vaccine Debate*, AJPH open-themed research, 108, 10, 1378-1384, 2018

Brown, O., Crawford, A., *Rising Temperatures, Rising Tensions - Climate change and the risk of violent conflict in the Middle East,* International Institute for Sustainable Development, 2009

Brosseder, U. B., *Xiongnu and Huns*, Archaeological Perspectives on a Centuries-Old Debate about Identity and Migration in Empires and Exchanges in Eurasian Late Antiquity Rome, China, Iran, and the Steppe, ca. 250-750, edited by Nicola di Cosmo, Institute for Advanced Study, Princeton, New Jersey, Michael Maas, Rice University, Houston, 2018

Brown, A., Pluskowski, A., *Medieval landscape transformation in the southeast and Eastern Baltic: paleoenvironmental perspectives on the colonisation of frontier landscapes*, Archaeologia Baltica 20, Frontier Societies and environmental change in northeast Europe, Klaipeda, 2013

Buchanan-Smith, M., *Darfuri migration from Sudan to Europe: from displacement to despair,* Humanitarian Policy group, 2018

Bungten, U., et al., 2500 Years of European Climate Variability and Human Susceptibility, Science, 311, 578-82, 2011

Bungten, U., et al. *Climate and societal change during the Late Antique Little Ice Age from 536 to 660,* Nature Geoscience, 9, 231-236, 2016

Burns, T., *Theodoric the Great and the concepts of power in late antiquity,* Acta Classica, 25, 99-118, 1982

Burki, T., *Vaccine misinformation and social media*, The Lancet, volume 1, 6, 258-259, 2019;

Burrows, K., Linney, P.K., *Exploring the Climate Change, Migration and Conflict Nexus*, Int. J. Environm. Res. Public Health, 13, 4, 443, 2016

Bury, J.B., *The Invasion of Europe by the Barbarians*, Jovian Press, ebook, 2017

Bustikova, L., Guasti, P., *The Illiberal Turn or Swerve in Central Europe?* Politics and Governance, 5, 4, 166-176, 2017

Cancello, R.; Soranna, D.; Zambra, G.; Zambon, A.; Invitti, C. *Determinants of the Lifestyle Changes during COVID-19 Pandemic in the Residents of Northern Italy*, Int. J. Environ. Res. Public Health, 17, 6287, 2020

Campion, E., *Ebola,* New England Journal of Medicine, 382, 1832-42, 2020

Castex, D., Kacki, S., *Demographic Patterns Distinctive of Epidemic Cemeteries in Archaeological Samples*. Microbiology Spectrum, American Society for Microbiology, 4, 4, 2016

Centres for Disease Control and Prevention, *First Global Estimates of 2009 H1N1 Pandemic Mortality Released by CDC-Led Collaboration*, 2012

Chen, C. et al., *University of Notre Dame Global Adaptation Index Country Index Technical Report,* ND-GAIN, 2015

Cicero, *On oratory and orators,* Watson, J.S., Harper & Brothers, New York, 1860

Ciscar, J., *Physical and economic consequences of climate change in Europe*, PNAS, 108, 7, 2678-2683, 2011

Cheeseman, N., *A divided Continent,* BTI, 2018

Cheyette, F., *The disappearance of the ancient landscape and the climatic anomaly of the early Middle Ages: a question to be pursued,* Early Medieval Europe, 16, 2, 127-165, 2011

Climate Change in the Breadbasket of the Roman Empire. Reconstructing Nile Floods for the Roman Period, Conference Report, Conference organized by the *Istituto Svizzero di Roma,* 23-24 January 2020

Climate Refugees: Beyond the Legal Impasse? Edited by Behrman, S. and Kent, A, Taylor and Francis, ebook, 2018

Coleman, D., *Does Europe Need Immigrants? Population and Work Force Projections,* The International Migration Review,26, 2, Special Issue: The New Europe and International Migration, 413-461, 1992

Colombeau, S.,Crisis of Schengen? *The effect of two 'migrant crises' (2011 and 2015) on the free movement of people at an internal Schengen border*, Journal of Ethnic and Migration Studies, 46, 11, 2258-2274, 2020

Compatangelo-Soussignan R., *Poseidonios and the original cause of the migration of the Cimbri: tsunami, storm surge or tides?* Revue des études anciennes, Revue des études anciennes, Université Bordeaux Montaigne, 2016

Cook, B., et al., *Spatiotemporal drought variability in the Mediterranean over the last 900 years,* J Geophys Res Atmos, 121, 5, 2060-2074, 2016

Cook, E. R., *Megadroughts, ENSO, and the Invasion of Late-Roman Europe by the Huns and Avars,* in The Ancient Mediterranean Environment between Science and History, Brill, Leiden, The Netherlands, 2013

Cole, J., Dodds, K., *Unhealthy geopolitics: can the response to COVID-19 reform climate change policy?*, Bull World Health Organ., 99, 2, 148-154, 2021

Colucci, M., *Storia dell'immigrazione straniera in Italia, dal 1945 ai giorni nostri,* Carocci Editore, Roma, 2017

COST Action 734, *Impact of climate change and variability on European agriculture'*, edited by Simone Orlandini, Pavol Nejedlik, Firenze University Press, Firenze, 2012

Crawford, M., *Famine and Disease in Ireland,* edited by A. Clarckson and Margaret Crawford, Taylor and Francis Ltd, London, 2005

Crawley, H, et al., *Destination Europe? Understanding the dynamics and drivers of Mediterranean migration in 2015,* Centre for Trust, Peace and Social Relations, Coventry University, Coventry, 2016

Dalby, S., *Climate Change and Geopolitics,* Oxford Encyclopaedia of Climate Science, Oxford University Press, online, 2017

Dall'Impero Romano a Carlo Magno, La Storia, La Biblioteca di Repubblica, UTET, Roma, 2004

Dark, K.R., *Large-scale population movements into and from Britain south of Hadrian's Wall in the fourth to sixth centuries AD,* Reading Medieval Studies, 29, 31-49, University of Reading, 2003

Darvas, Z., *COVID-19 has widened the income gap in Europe,* Bruegel Blog, 3rd of December 2020

Daszak, P., et al., *A strategy to prevent future epidemics similar to the 2019-nCoV outbreak,* Biosafety and Health, 2,1, 6-8, 2020

Davis, E.D., *The Scarcity of Rats and the Black Death*: An Ecological History, The Journal of Interdisciplinary History, 16, 3, 455-470, 1986

De Jong, R., Lageras, P., *Exploring the patterns and causes of land use changes in south-west Sweden,* Veget Hist Archaeobot, 20, 15-27, 2011

De Wit, E., et al., *SARS and MERS: recent insights into emerging coronaviruses,* Nature reviews, microbiology, 14, 523-534, 2016

del Rio C, Guarner J., *The 2009 influenza A (H1N1) pandemic: what have we learned in the past 6 months,* Trans Am Clin Climatol Assoc., 121, 128-140, 2010

De Torre, C., *Routledge Handbook of Global Populism,* Taylor and Francis, ebook, 2018

Desprat, S., et al., *Revealing climatic variability of the last three millennia in north-western Iberia using pollen influx data,* Earth and Planetary Science Letters, 213, 63-78, 2003

Devakumar, D., et al., *Racism and discrimination in COVID-19 responses,* The Lancet, 395, 10231, 1194, 2020

Dimitriadi, A., *Migration from Afghanistan to third countries and Greece,* Hellenic Foundation for European and foreign policy, 2013

Dockery, W., *What's behind the influx of Bangladeshi migrants to Europe?* InfoMigrants, 14th of July 2017

Dong, E., Du, H., Gardner L. *An interactive web-based dashboard to track COVID-19 in real time,* Lancet Inf Dis. 20,5, 533-534, 2020

Dolukhanov P. *The Early Slavs: Eastern Europe from the Initial Settlement to the Kievan Rus,* Taylor and Francis, ebook, 2014

Drake, B., Lee, *Changes in North Atlantic Oscillation drove Population Migrations and the Collapse of the Western Roman Empire,* Sci Rep, 7, 1227, 2017

Duff, A., Rankin, S., *Exploring flexible home arrangements - an interview study of workers who live in vans,* Career Development International, 25, 7, 747-761

Duroselle, J., *L'invasion, Les migration humaines, chance ou fatalité?* PLON, Paris, 1992

EASO, *Annual Report on the Situation of Asylum in the European Union,* 2020

Eckert, E., *Sustainability in the European Union: Analyzing the Discourse of the European Green Deal,* Journal of Risk and Financial Management, MDPI, 14, 2, 80, 2021

Elmqvist, T., et al., *Urbanization, habitat loss and biodiversity decline, solution pathways to break the cycle,* In: K.C. Seto, W.D. Solecki, C.A. Griffith (Eds.), The Routledge Handbook of Urbanization and Global Environmental Change. Routledge, Abingdon, Oxon, UK, 139-151, 2016

Elverhøi, Y. A., *Germanic piracy in Roman Britain*, A Thesis Presented to The Department of Literature, Area Studies and European Languages, University of Oslo, 2010

Elkerbout, M., The European Green Deal after Corona: Implications for EU climate policy, CEPS, 6, 2020

Eroshenko GA, Nosov NY, Krasnov YM, Oglodin YG, Kukleva LM, Guseva NP, et al., *Yersinia pestis strains of ancient phylogenetic branch 0.ANT are widely spread in the highmountain plague foci of Kyrgyzstan*, PLoS ONE 12, 10, 2017

Esper, J., et al., *Northern European summer temperature variations over the Common Era from integrated tree-ring density records*, Journal of Quaternary Science, 29, 5, 487-494, 2014

Evagrius, *Ecclesiastic History, History Of The Church, In Six Books, From A. D. 431 To A.D. 594, A New Translation From The Greek, with an account of the author and his writings,* Samuel Bagster And Sons, London, 1846

European Union, *2021 report on gender inequality in the EU,* 2021

Ezeh, A., et al., *Why sub-Saharan Africa might exceed its projected population size by 2100,* 396, 10258, 1131-1133, 2020

Fahad, S., Wang, S., *Climate change, vulnerability, and its impacts in rural Pakistan: a review,* Environmental Science and Pollution Research, 27, 1334-1338, 2020

Falkenberg, L. et al., *Ocean acidification and human health,*Int J Environ Res Public Health, 17, 12, 4563, 2020

Fawzy, S., et al., *Strategies for mitigation of climate change: a review,* Environmental Chemistry Letters, 18:2069-2094, 2020

Feitelson, E., Tubi, A., *A main driver or an intermediate variable? Climate change, water and security in the Middle East*, Global Environmental Change, 44, 39-48, 2017

Feng, D. et al., *The SARS epidemic in mainland China: bringing together all epidemiological data,* Tropical medicine & international health, 14,1, 4-13. 2009

Fernandez-Sebastian M., *Maritime Dimension in the Fight Against Illegal Migration on the Western Mediterranean Route,* in: Ramírez, J., Biziewski J. (eds) Security and Defence in Europe. Advanced Sciences and Technologies for Security Applications. Springer, Cham, 2020

Flink, G., *Ölands stensträngsområden och den justinianska pesten*, Nordic Journal of Settlement History and Built Heritage, 11, 21-135, 1986

Forte, A., Oram, D., et al., *Viking Empires,* Cambridge University Press, Cambridge, 2005

Freeman,L. *Environmental Change, Migration, and Conflict in Africa: A Critical Examination of the Interconnections*, Journal of Environment & Development 26,4, 351-374, 2017

Furber, M., Evengard B., Nilsson, M., *Facing the limit of resilience: perceptions of climate change among reindeer herding Sami in Sweden*, Global Health Action, 2019

Gage, K.L., et al., *Climate and vectorborne diseases*, American Journal of Preventive Medicine, 35,5, 436-50, 2008

Gaidhane, S., et al., *Depression, anxiety and stress among the general population in the time of COVID-19 lockdown: A cross-sectional study protocol*, International Journal of Research in Pharmaceutical Sciences, 11, 1, 360-364, 2020

Ganshof, F., *Feudalism,* University of Toronto Press, Medieval Academy of America, 1996

Gauci, A. et al., *The 2009 A(H1N1) pandemic in Europe, A review of the experience*, Special Report, ECDC, 2010

Gaillard, M., et al., Causes of Regional Change-Land Cover, Second assestment of climate change for the Baltic Basin, The BACC II Author team, Springer Open, Regional Climate Studies, 2015

Gebre Selassie, Y., *Plague as a Possible Factor for the Decline and Collapse of the Aksumite Empire: a New Interpretation*, Northeast African Journal of Social Sciences and Humanities, 1, 2011

Gentili, A., *Il leone e il cacciatore,* Carocci Editore, Roma, 2015

Gerhards, J., et al., *How Strong is European Solidarity?* Berlin Studies on the Sociology of Europe (BSSE), 37, 2018

Gildas, *The ruin of Britain, fragments from lost letters, the Penitential, Together with The lorica of Gidas*. Edited for the Hon. Society of Cymmrodorion by Hugh Williams, Published by the the Hon. Society of Cymmrodorion by David Nutt, London, 1899

Gleick, P., *Water, Drought, Climate Change, and Conflict in Syria,* Weather, Climate, and Society, 6, 3, 331-340, 2014

Goffart, W.A., *Barbarians and Romans, A.D. 418-584: The Techniques of Accommodation*, Princeton University Press, Princeton, 2020

Gouvea Dos Santos, W., *Natural history of COVID-19 and current knowledge on treatment therapeutic options*, Biomedicine & Pharmacotherapy, 129, 110493, 2020

Graslun, B., Price, N., *Twilight of the gods? The 'dust veil event' of AD 536 in critical perspective*, Antiquity, 86(332), 428-443, 2012

Grinsted, A., et al, *Sea level rise projections for northern Europe under RCP8.5,* Clim Res, Vol. 64, 15-23, 2015

Guasti, P., *The Impact of the COVID-19 Pandemic in Central and Eastern Europe,* Democratic Theory, 7, 2, 47-60, 2020

Guinotte, J., Fabry, V., *Ocean Acidification and Its Potential Effects on Marine Ecosystems, New York Academy of Sciences,* 1134, 320-342, 2008

Gundersen, I.M., *The Fimbulwinter theory and the 6th century crisis in the light of Norwegian archaeology: Towards a human-environmental approach,* Primitive Tider 21, 101-120, 2017

Güneralp, B., *Urbanization in Africa: challenges and opportunities for conservation,* Environmental Research Letters, 13, 1, 2017

Gunn, J.D., *The years without summer: tracing AD 536 and its aftermath,* British Archaeological Reports International, Archaeopress, Oxford, 2000

Gwynn D.M., *The Goths: Lost Civilizations*, Reaktion Books, London, 2017

Haamed, M., *Drought and food security in the middle east: An analytical framework,* Agricultural and Forest Meteorology, 281, 107816, 2020

Haldon, J., et al., *Plagues, climate change, and the end of an empire: A response to Kyle Harper's The Fate of Rome (1):* Climate, History Compass, Wiley Online Library, 2018

Haley, G.C., *Tamlachta: The Map of Plague Burials and Some Implications for Early Irish History*, Proceedings of the Harvard Celtic Colloquium, 22, 96-140, 2002

Halsall, G., *Barbarian Migrations and the Roman West, 376-568,* Cambridge University Press, Cambridge, 2007

Hakenbeck, S., *The Role of Climatic and Environmental Factors in the Hunnic Phenomenon in South-east Europe*, SocArXiv, 2019

Hassoun, A., et al., *Acidification of the Mediterranean Sea from anthropogenic carbon penetration, Deep Sea Research Part I: Oceanographic Research Papers,* 102, 1-15, 2015

Harbeck M, Seifert L, Hänsch S, Wagner DM, Birdsell D, Parise KL, et al., *Yersinia pestis DNA from Skeletal Remains from the 6th Century AD Reveals Insights into Justinianic Plague*, PLoS Pathog 9, 5, 2013

Harke, H., *Anglo-Saxon Immigration and Ethnogenesis*, Medieval Archaeology, 55, 1-28, 2011

Harmatta, J., *Annexation of the Hephthalite vassal kingdoms by the western Turks* in History of Humanity: From the seventh century B.C. to the seventh century A.D, edited by Joachim Herrmann and Erik Zurker, UNESCO, 1996

Harper, K., *Invisible Environmental History: Infectious Disease in Late Antiquity*, Late Antique Archaeology, 12, 1, 116-131, 2016

Harper, K., *The Fate of Rome, Climate, Disease and the End of an Empire,* The Princeton History of the Ancient World, Princeton University Press, Princeton, 2017

Harrison, D., *Plague, Settlement and Structural Change at the Dawn of the Middle Ages,* Scandia, Vol 59,1, 15-49,1993

Harrison, D., *Stora Doden,* Ordfront, Stockholm, 2000

Hawkes, S.C., *Anglo-Saxon Kent c 425-725*, in Archaeology in Kent to AD 1550: in memory of Stuart Eborall Rigold, Edited by D.G., Buckley, CBA Research Report, 1982

Helama S., et al., *Dark Ages Cold Period: A literature review and directions for future research,* The Holocene, Volume 27, 1600-1606, 2017

Helama et al., *Limited Ice Age Cooling,* Nature Geoscience, 10, 242-3, 2017

Helama, S., et al., *Volcanic dust veils from sixth century tree-ring isotopes linked to reduced irradiance, primary production and human health*, Scientific reports, Nature, 1339, 2018

Hennigan, W., *Lost in the Pandemic: Inside New York City's Mass Graveyard on Hart Island,* 18th November 2020

Heldring, L., Robinson, J., *Colonialism and development in Africa,* The Oxford Handbook of the Politics of Development, Edited by Carol Lancaster and Nicolas van de Walle, 2015

Herring, D., *Isn't there a lot of disagreement among climate scientists about global warming?* NOAA, Climate.gov, 3rd of February 2020

Heather, P., *La caduta dell'Impero romano: una nuova storia*, Garzanti, Milano, 2006

Heather, P., *The Huns and the End of the Roman Empire in Western Europe*, English Historical Review,110, 435, 4-42, 1995

Hien, L., et al., *The positive impact of lockdown in Wuhan on containing the COVID-19 outbreak in China*, Journal of Travel Medicine, 27, 3, 32181488, 2020

Hirschfeld, Y., *The crisis of the sixth century: climatic change, natural disasters and the plague*, Mediterranean Archaeology and Archaeometry, 6, 1, pp. 19-32, 2006

Hooghe, M., et al., *Migration to European countries: a structural explanation of patterns, 1890-2004,*

IMR: International Migration Review, 42, 2, 476-504, 2008

Holt, D.H., *Germania and Climate Variability in 3rd and 4th Centuries A.D.: A Methodological Approach to Dendroclimatology and Human Migration*, Physical Geography, 32, 3, 241-268, 2011

Howorth, H.H., *The Avars, The Journal of the Royal Asiatic Society of Great Britain and Ireland, Oct., New Series, 21, 4, 721-810, 1899*

Hsiu-Ching Ho, M., Liu, J., *The swift knowledge development path of COVID-19 research: the first 150 days*, Scientometrics, 126, 2391-2399, 2021

Huang, C., et al., *Clinical features of patients infected with 2019 novel coronavirus in Wuhan, China,* Lancet, 395, 10223, 497-506, 2020

Hufthammer, A.K.,Walløe, L., Rats cannot have been intermediate hosts for Yersinia pestis during medieval plague epidemics in Northern Europe, Journal of Archaeological Science, Elsevier, Volume 40, 4,1752-1759, 2013

Humphreys, M., *Economics and violent conflicts,* Harvard University, Harvard, 2003

Huntjiens, P., Nachbar, K., *Climate Change as a Threat Multiplier for Human Disaster and Conflict,* Working paper 9, The Hague Insititute for Global Justice, 2015

Hurrell, J.W., et al., *The North Atlantic Oscillation: Climatic Significance and Environmental Impact*, American Geophysical Union as part of the Geophysical Monograph Series, 134, 73-73, 2013

Idemudia, E., *Social experiences of Migrants,* in Psychosocial Experiences of African Migrants in Six European Countries, pp 119-135, 2020

Idemudia, E., *Theoretical Explanations of Migrations, Mental Health, Wellbeing and Posttraumatic Stress Disorder,* in: Psychosocial Experiences of African Migrants in Six European Countries. Social Indicators Research Series, vol 81. Springer, Cham, 2021

Idemudia, E., *Travelling Routes to Europe,* in Idemudia, E., Boehnke, K., *Psychosocial Experiences of African Migrants in Six European Countries*, Social indicators Research Series, Springer, 33-49, 2020

IDMC, *Global Report on Internal Displacement 2021,* 2021

Il Canzoniere eddico, a cura di Piergiuseppe Scardigli, Garzanti Editore, Milano, 2004

Internal Displacement Monitoring Center, *Recommendations for addressing drought displacement in Ethiopia*, Briefing Paper, 2021

International Federation of Red Cross and Red Crescent Societies, *Drought in the Horn of Africa. Preventing the next disaster,* 2011

IOM, *Current migration trends from Bangladesh to Italy Focus on arrivals by sea,* IOM Italy Briefing, 1, June 2017

IOM, *Irregular migration towards Europe, Western African Route: Migration to the Canary Island,* 18 November 2020

IOM, Mixed migration flows in the Mediterranean, 2020

IOM, Pakistan: survey on drivers of migration, 2020

IOM, *Sudan 2017-9*, Migration crisis operational framework, 2019

Iliffe J., *The African AIDS Epidemic: A History*, Ohio University Press, Athens, Ohio, USA, 2005

James, E., *Europe's Barbarians AD 200-600*, Taylor and Francis, ebook, 2014

Jakkola, J., *The Holistic Effects of Climate Change on the Culture, Well-Being, and Health of the Saami, the Only Indigenous People in the European Union,* Curr Environ Health Rep.,5,4, 401-417, 2018

Jankuhn, H., *Trade and Settlement in Central and Northern Europe up to and during the Viking Period*, The Journal of the Royal Society of Antiquaries of Ireland, 112, 18-50, 1982

Jeffries, B., *The loss of and the rise of Nature pandemics*, WWF, 2020

Jellema, D., *Frisian trade in the Dark Ages, Speculum,* The University of Chicago Press, 30, 1, 15-36, 1955

Johnson, S., *Later Roman Britain*, Taylor and Francis, ebook, 2014

Johnstone, S., Mazo, J., *Global Warming and the Arab Spring,* in The Arab Spring and Climate Change, A Climate and Security Correlations Series, Edited by Caitlin E. Werrell and Francesco Femia, Center for American Progress, Stimson, The center for Climate and Security, 2013

Jones, M.U., *Mucking and Early Saxon rural settlement in Essex,* in Archaeology in Essex to AD 1550, Edited by D.G., Buckley, CBA Research Report, The Council for British Archeology, London, 34, 1980

Jones, M., *The end of Roman Britain,* Cornell University Press, Ithaca, New York, 1998

Jones, A.H.M., *The decline and fall of the Roman Empire,* History, 40, 140, 209-226, 2005

Jordanes, *The origin and the deeds of the Goths,* translated by Charles C. Mierow, CreateSpace Independent Publishing Platform, 2017

Josiah C., *Population in Europe* in: Carlo M. Cipolla, ed., The Fontana Economic History of Europe, Vol. I: The Middle Ages, 25-71, 1972

Kang, Y., *Refugee crisis in Europe: determinants of asylum seeking in European countries from 2008-2014*, Journal of European Integration, 43, 1, 2021

Kazanski, M., *Archaeology of the Slavic Migrations*, Encyclopedia of Slavic Languages and Linguistics Online, 2020

Keller, M., Spyrou, M.A., et al., *Ancient Yersinia pestis genomes from across Western Europe reveal early diversification during the First Pandemic (541-750)*, PNAS, 116, 25, 12363-12372, 2019

Kelley, C. et al., *Climate change in the Fertile Crescent and implications of the recent Syrian drought*, PNAS, 112, 11, 3241-3246, 2015

Kelman, I., Naess, M. W., *Climate Change and Displacement for Indigenous Communities in Arctic Scandinavia*, Center for International Climate and Environmental Research, Oslo, 2013

Kelman, I., Naess, M. W., *Climate Change and Migration for Scandinavian Saami: A Review of Possible Impacts*, MDPI, 7, 47, 2019

Kibreab, G., *The national service/Warsai-Yikealo Development Campaign and forced migration in post-independence Eritrea*, Journal of Eastern African Studies, 7,4, 630-649, 2013

Kim, H.J., *The Huns*, Taylor and Francis, ebook, 2015

Kindhauser, M.K., *Zika: the origin and spread of a mosquito-borne virus*, Bulletin of the World Health Organization, 94, 675-686, 2016

Kirwin, M., Anderson, J., *Identifying the factors driving west African migration*, OECD, West African Papers, 17

Knopf, P., *South Sudan's Civil War and Conflict Dynamics in the Red Sea*, Special report, United States Institute of Peace, 2018

Kraler, A., et al., *Climate Change and Migration: Legal and policy challenges and responses to environmentally induced migration*, Policy Department for Citizens' Rights and Constitutional Affairs Directorate-General for Internal Policies, European Parliament, 2020

Kolby, J., *To prevent the next pandemic, it's the legal wildlife trade we should worry about*, National Geographic, 7th of May 2020

Kovács, Z.E., *Dispersal history of an invasive rodent in Hungary - subfossil finds of Rattus Rattus*, Acta Zoologica Academiae Scientiarum Hungaricae 58, 4, 379-394, 2012

Kundera, J., *The Future of EU: Towards a two Speed Europe*, European Research Studies Journal, 22, 3, 261-281, 2019

Kurbanov, A., The Hephthalites: archaeological and historical analysis, Berlin Freie Universität, Berlin, 2010

Kushminder, K., et al., *Irregular Migration Routes to Europe and Factors Influencing Migrants' Destination Choices*, Maastricht Graduate School of Governance, 2015

Larsen, L.B., *New ice core evidence for a volcanic cause of the A.D. 536 dust veil*, Geophysical Research Letters, 35, 4, 2008

Le Storie di Ammiano Marcellino, tradotte da Francesco Ambrosoli, Antonio Fontana Editore Milano, 1829

Lendaro, A., *A 'European Migrant Crisis'? Some Thoughts on Mediterranean Borders*, Studies in ethnicity and Nationalism, 16, 1, 148-157, 2016

Lee, D., *The Role of Hostages in Roman Diplomacy with Sasanian Persia*, Historia: Zeitschrift für Alte Geschichte, 40, 366-374, 1991

Lindsey, P., *The bushmeat trade in African savannas: Impacts, drivers, and possible solutions*, Biological Conservation, 160, 80-96, 2013

Little, K.L. (ed), Plague and the end of antiquity, the pandemic of 541-750, Cambridge University Press in association with The American Academy in Rome, New York, 2007

Liu, Y., et al., *COVID-19, the first documented coronavirus pandemic in history*, Biomedical Journal, 43, 328-333, 2020

Lopez, I., *La Unión Europea como factor de paz*, Agenda internacional, XI, 22, 25-53, 2005

Lopez-Bueno, R., et al., *Potential health-related behaviors for pre-school and school-aged children during COVID-19 lockdown: A narrative review*, Preventive Medicine, 143, 106349, 2021

Lopreite, M., *Early warnings of COVID-19 outbreaks across Europe from social media*, Scientific Reports, 11, 2147, 2021

Lovei, M., *Climate Impacts on African Fisheries: The Imperative to Understand and Act*, World Bank Blog, 11th of November 2017

Löwenborg, D. *An Iron Age Shock Doctrine - Did the AD 536-7 event trigger large-scale social changes in the Mälaren valley area?*, Journal of Archaeology and Ancient History, 4, 1-29, 2012

Luppi, F., Arpino, B., Rosina, A., *The impact of COVID-19 on fertility plans in Italy, Germany, France, Spain, and the United Kingdom*, Demographic Research, 43, 47, 1399-1412, 2020

Lutherbach et al., *European summer temperatures since Roman times*, Environmental Research Letters, 11, 2016

Lynn, C.J., *Settlement and Disease: a plague on your raths* in *Archaeology Ireland*, 19,4, 14-7, 2005

MacArthur, W., *The Identification of Some Pestilences Recorded in the Irish Annals*, Irish Historical Studies, 6, 23, pp. 169-188, 1949

Macbain, B., *Odovacer the Hun?*, Classical Philology, 78, 4, 323-327, The University of Chicago Press, 1983

MacDowall, S., *The Franks*, Pen and Sword, Barnsley, UK, 2018

Machacek, J., *Runes from L'any (Czech Republic) - The oldest inscription among Slavs. A new standard for multidisciplinary analysis of runic bones*, Journal of Archaeological Science, London, Academic Press, 2021, vol. 127, N. 1, p. 1-8127, 2021

Marc, A., *The Challenge of Stability and Security in West Africa,* International Bank for Reconstruction and Development / The World Bank, 2015

Marois, G., Potančoková, M., *Scenarios of labour force participation and employment integration of immigrants in the EU: demographic perspective*, JRC Technical report, European Commission, 2020

Mbiyozo, A., *Fleeing terror, fighting terror: the truth about refugees and violent extremism,* Institute for Security Studies, 2018

McCormick, M., *Climate Change during and after the Roman Empire: Reconstructing the Past from Scientific and Historical Evidence*, Journal of Interdisciplinary History, 43, 2, 169-220, 2012

McCormick, M., *Tracking mass death during the fall of Rome's empire (I). Journal of Roman Archaeology*, 28, 325-357, 2015

McDonald, R., et al., The Future of Global Urbanization and the Environment, Solutions, 2016

McDonald, R., et al., *The implications of current and future urbanization for global protected areas and biodiversity conservation,* Biological Conservation, 141, 6, 1695-1703, 2008

McMichael, A.J., *Paleoclimate and bubonic plague: a forewarning of future risk?* BMC Biology 8, 108, 2010

Malgesini, G., *The impact of Covid-19 on people experiencing poverty and vulnerability,* European Anti-poverty network report, July 2020

Mahmoeei, B., Parris, B., *Climate change and internal migration patterns in Bangladesh: an agent-based model,* Environment and Development Economics, 17,6, 763-780, 2012

Marchand, K., *Study on Migration Routes in the East and Horn of Africa*, Maastricht Graduate School of Governance, 2017

Maroukis, T., & Gemi, E. *Albanian circular migration in Greece: beyond the state?* In A. Triandafyllidou (Ed.), Circular Migration between Europe and its Neighborhood: Choice or Necessity? Oxford University Press, 68-89, 2013

Martinez, V., et al., The assessment of lifestyle changes during the COVID-19 pandemic using a multidimensional scale, Revista de Psiquiatría y Salud Mental, 14, 1, 16-26, 2021

Marzocchi, O., *The impact of Covid-19 measures on Democracy, the Rule of Law and Fundamental rights, Policy Department for Citizens' Rights and Constitutional Affairs,* Briefing requested by the LIBE committee, Monitoring Group on Democracy, Rule of Law, Fundamental Rights, 2020

Mastrojeni, G., Pasini, F., *Effetto guerra effetto serra,* Chiarelettere, Milano 2017

Mathisen, R., *"Becoming Roman, Becoming Barbarian": Roman Citizenship and the Assimilation of Barbarians into the Late Roman World,* Migration and Membership Regimes in Global and Historical Perspective, Studies in Global Social History, 13, 2, 2013

Mathisen, R., *Peregrini, Barbari, and Cives Romani: Concepts of Citizenship and the Legal Identity of Barbarians in the Later Roman Empire*, The American Historical Review, 111, N. 4, 1011-1040, 2006

Cardona, M, et al., *The macro-economic impact of e-commerce in the EU Digital Single Market,* JRC Technical reports, European Commission, 2015

Meier, M., *The 'Justinianic Plague': An "Inconsequential Pandemic"? A Reply,* Medizinhistorisches Journal, Volume 55, 2, 172-199, 2020

Meir, M., *The 'Justinianic Plague': the economic consequences of the pandemic in the eastern Roman empire and its cultural and religious effects,* Early Medieval Europe, 24,3, pp. 267-292, 2016

Meeson Beck, F.G., *Alemanni,* Encyclopaedia Britannica, Volume 1, London, 1911

Merrills A, *Vandals, Romans and Berbers: New Perspectives on Late Antique North* Africa, Taylor and Francis, ebook, 2017

Middleton, G.D., *Understanding Collapse: Ancient History and Modern Myths*, Cambridge University Press, Cambridge, 2018

Miller, T., *The Birth of the Hospital in the Byzantine Empire,* Johns Hopkins University Press, Baltimora, 1997

Momin, S. *Human Rights Based Approach to Refugees: Look at the Syrian Refugee Crisis and the Responses from Germany and the United States,* Duke Forum for Law and Social Change, 9, 55-80, 2017

Mordechai, L., Eisenberg, M., *Rejecting Catastrophe: The Case of the Justinianic Plague*, Past & Present, Volume 244, Issue 1, August, 3-50, 2019

Morrison, T., *Britain and the Anglo-Saxons in Late Antiquity,* University of Mexico Digital Repository, 2016

Moschen, R., et al., *Temperature variability at Dürres Maar, Germany during the migration period and at high medieval times, inferred from stable carbon isotopes of Sphagnum cellulose,* Clim. Past Discuss., 7, 535-573, 2011

Mukthar, S., et al., *Boko Haram and the Geopolitics of Forced Migration in Nigeria,* Journal of International Studies Vol. 14, 51-63, 2018

Munsch, T., et at., *Before the desert,* Conditions and Risks on Mixed Migration Routes through West Africa, Danish Refugee Council, 2017

Myres, J. N. L., *The Anglo-Saxon Settlements,* Oxford University Press, Oxford 1969

Nae, T., Panie, N., *European Green Deal: The Recovery Strategy Addressing Inequalities*, Journal of Eastern Europe Research in Business and Economics, 2021

Naguib, N., Humanitarian Pluralism: *The Arctic Passage in an Age of Refugees*, International Journal of Middle East Studies, Cambridge University Press, 377-381, 2016

Napier, W., *The Memory Remains: Why the Migration Period and the Fall of Rome Continue to be Mischaracterized as a Barbarian Invasion,* Electronic Theses and Dissertations, 2020

Neparáczk, E., et al., *Y-chromosome haplogroups from Hun, Avar and conquering Hungarian period nomadic people of the Carpathian Basin*, Scientific Reports, Sci.Rep.9,1, 16569, 2019

Nerem, R. et al., *Climate-change-driven accelerated sea-level rise detected in the altimeter era,* PNAS February 27, 115, 9, 2022-2025, 2018

Neukom, R., No evidence for globally coherent warm and cold periods over the pre-industrial Common Era, Nature, 571, 550-554, 2019

Newfield, T.P., *Mysterious and mortiferous clouds, The Climate Cooling and Disease Burden of Late Antiquity*, Late Antique Archaeology 12, 1, 89-115, 2016

Ngondjeb, D., et al., *Insights on Africa's Future in its Transition to the Green Economy,* In: Atewamba C., Yong Ngondjeb D. (eds) Inclusive Green Growth. Advances in African Economic, Social and Political Development. Springer, Cham, 2020

Nieto-Moreno, V., et al., *Tracking climate variability in the western Mediterranean during the Late Holocene: a multiproxy approach*, Climate of the Past, 7, 1395-1414, 2011

Nyong, A., et al., *The value of indigenous knowledge in climate change mitigation and adaptation strategies in the African Sahel,* Mitigation and Adaptation Strategies for Global Change, 12, 787-797, 2007

Onya, C., *Climate Change and Conflict in Nigeria: The Boko Haram Challenge,* American International Journal of Social Science, 4, 2, 181-190, 2015

Osterholm, M., Olshaker, M., Chronicle of a Pandemic Foretold, Foreign Affairs, 99, 4, 10-25, 2020

Palamenghi, L., et al., *Mistrust in biomedical research and vaccine hesitancy: the forefront challenge in the battle against COVID-19 in Italy*, Eur J Epidemiol 35, 785-788, 2020

Palomino, J., et al., *Wage Inequality and poverty effects of the lockdown in Europe*. European Economic Review, 129, 103564, 2020

Paolo Diacono, *Historia Langobardorum,* Foro Barbarico, GBL Grande Biblioteca, ebook, 2020

Pasini, A., Mazzocchi, F., Perception and risk of Covid-19 and climate change: Investigating analogies in a common framework. Global Sustainability, 3, E32, 2020

Patrick, S., *When the system fails,* Foreign Affairs, 99, 4, 40-51, 2020

Pauchard, A., *Multiple effects of urbanization on the biodiversity of developing countries: The case of a fast-growing metropolitan area (Concepción, Chile),* Biological Conservation, 127, 3, 272-281, 2006

Peixoto, J., et al., *Immigrants, markets and policies in Southern Europe,* in Immigrants, markets and policies in Southern Europe, Trends, Structures and Policy Implications, edited by Marek Okólski, IMISCOE Research, Amsterdam University Press, 2012

Peregrine, P., *Climate and social change at the start of the Late Antique Little Ice Age*, The Holocene, 30, 11, 1643-8, 2020

Petersen, L.R., et al., *Zika Virus*, New England Journal of Medicine, 374, 1552-1563, 2016

Piganiol, A. *The causes of the fall of the Roman Empire*, The Journal of General Education, 5, 1, 62-69, 1950

Pohl W., *The Avars: A Steppe Empire in Central Europe, 567-822*, Cornell University Press, Ithaca, New York, USA, 2018

Portail sur les données migratoires, *Migration data in Eastern Africa,* 02 February 2021

Powell, J., *Scientists Reach 100% Consensus on Anthropogenic Global Warming,* Bulletin of

Science, Technology & Society, 37, 4,183-184, 2017

Price N. and B. Gräslund B., *Excavating the Fimbulwinter? Archaeology, geomythology and the climate event(s) of AD 536* In: Past Vulnerability. Volcanic eruptions and human vulnerability in traditional societies past and present. Riede F., 109-132. Aarhus University Press, Aarhus, 2015

Price, R., *Climate change as a driver of conflict in Afghanistan and other Fragile and Conflict Affected States,* K4D Helpdesk Report 527, Institute of Development Studies, Brighton, UK, 2019

Prívara, A., Prívarová, M., Nexus between Climate Change, Displacement and Conflict: Afghanistan Case, Sustainability, 11, 5586, 2019

Procopius, *History of the Wars,* translated by H. B. Dewing, Loeb Classical Library 48, Harvard University Press, Cambridge, 1914.

Procopius, History of the Vandalic Wars Volume 2, Books III and IV, William Heinemann, London, England 1916

Procopius, *Secret History,* translated by Richard Atwater, University of Michigan Press, Ann Arbor, 1963

Pseudo-Dionysius of 'Tell-Mahre', *Chronicles,* Liverpool University Press, Liverpool, 1996

Pullano, G., et al., *Novel coronavirus (2019-nCoV) early-stage importation risk to Europe,* Euro Surveill., 25,4, 2020

Rahman, M., Kabir, M., *Moving to Europe: Bangladeshi Migration to Italy,* Institute of South Asian Studies Working Paper, 142, 2012

Rance, P., *Attacotti, Déisi and Magnus Maximus: The Case for Irish Federates in Late Roman Britain,* Britannia, Society for the Promotion of Roman Studies, 32, 243-270, 2001

Rance, P., *Epiphanius of Salamis and the Scotti: New Evidence for Late Roman-Irish Relations, Britannia,* The Society for the Promotion of Roman Studies, 43, 227-242, 2012

Ratovonjato, J., Rajerison, M., Rahelinirina, S., Boyer, S., *Yersinia pestis in Pulex irritans Fleas during Plague Outbreak,* Madagascar, Emerg Infect Dis., 20,8, 1414-1415, 2014

Rascovan, N., Sjögren, K., et al., *Emergence and Spread of Basal Lineages of Yersinia pestis during the Neolithic decline,* Cell, Volume 176, 1-2, 295-305, 2019

Ravegnani, G., *La caduta dell'Impero Romano,* Il Mulino, Bologna 2012

REACH, *Migration from Afghanistan to Europe (2014-17),* Research report, October 2017

Renda, A., Castro, R., *Towards Stronger EU Governance of Health Threats,* European Journal of Risk Regulation, 11, 2, 273-282, 2020

Retief, F.P., *The epidemic of Justinian (AD 542): a prelude to the Middle Ages,* Acta Theologica, Supplementum 7, 115-127, 2005

Ricci, C., *The Necessity for Alternative Legal Pathways: The Best Practice of Humanitarian Corridors Opened by Private Sponsors in Italy,* German Law Journal, Cambridge University Press, 21, 265-283, 2020

Riley, D., Emigh, R., Post-Colonial Journeys: Historical Roots of Immigration and Integration, Comparative Sociology, 1, 2, 169-193, 2002

Rios, A., et al., *Decadal acidification in the water masses of the Atlantic Ocean,* PNAS August 11, 112, 32, 9950-9955, 2015

Risse, Guenter B., *Mending Bodies, Saving Souls: A History of Hospitals.* Oxford University Press, Oxford, 1999

Robinson, J.H., *Readings in European History,* 35-39, Boston, New York, Ginn and co., 1905

Roche, B., et al., *Was the COVID-19 pandemic avoidable? A call for a "solution-oriented" approach in pathogen evolutionary ecology to prevent future outbreaks,* Ecology Letters, 23, 11, 1557-1560, 2020

Rogers, K., Dude, A., Siliciano, R., Dorwick, K., *AIDS,* Encyclopedia Britannica, 2020

Rogers, K, "*MERS*", Encyclopedia Britannica, 2017

Roksandic, *S., et al., Migration in the Time of COVID-19—Policy Responses and Practices in Croatia Concerning the Western Balkan Routes and Readiness for the Post-COVID-19 Society in Which the Right to Health Care for the Most Vulnerable Is Guaranteed,* Front. Hum. Dyn., 2021

Rolando, F., Naso, P, *Humanitarian Corridors to Italy: An Interview With Professor Paolo Naso,* Harvard International Review, 39,2, 64-67, 2018

Roos, R., *MERS outbreaks grow; Malaysian case had camel link,* Center for infectious Disease Research and Policy, University of Minnesota, 2014

Rosen, W., Justinian's Flea, Plague, Empire, and the Birth of Europe, Viking Penguin, New York, 2007

Ruffolo, G., *Quando l'Italia era una superpotenza,* Einaudi, Torino, 2004

Sabine, C., et al., *The Oceanic Sink for Anthropogenic CO_2,* Science, 305, 5682, 367-371, 2004

Sales-Carbonell, J., *Reaccions i estratègies davant l'arribada de la «pesta de Justinià» a la Península Ibèrica (segles VI-VII),* Revista d'igualada, 64-65, 2020

Saha, S., How climate change could exacerbate conflict in the Middle East, Atlantic Council, 14th of May 2019

Sanchez-Lopez, G., *Climate reconstruction for the last two millennia in central Iberia: The role of East Atlantic (EA), North Atlantic Oscillation (NAO) and their interplay over the Iberian Peninsula*, Quaternany Science Reviews 149, 135-150, 2016

Šantić, D., *The Balkan Migration Route: Reflections from a Serbian Observatory* in Towards understanding of Contemporary Migration Causes, Consequences, Policies, Reflections, edited by Mirjana Bobić, Stefan Janković, Institute for Sociological Research, Belgrade, 221-241, 2017

"SARS", Encyclopedia Britannica online

Search for Common Ground, *Trust in authorities – the golden ticket to successful COVID-19 vaccine roll-out in conflict settings*, January 2020

Sellar, A.M., *Bede's Ecclesiastical History of England, A Revised Translation, With Introduction, Life, and Notes*, London George Bell and Sons, Oxford, 1907

Schamiloglu, U., *The Plague in the Time of Justinian and Central Eurasian History: An Agenda for Research* in O. Karatay, & I. Zimonyi (Eds.), Central Eurasia in the Middle Ages: Studies in Honour of Peter B. Golden, Turcologica, 104, Harrassowitz Verlag, Wiesbaden, 2016

Schoorl, J., et al., *Push and pull factors of international migration A comparative report*, European Commission, 2009

Seger, T., *The plague of Justinian and other scourges : an analysis of the anomalies in the development of the iron age population in Finland*, Journal of Swedish Antiquities, 77, 184-197, 1982

Shahraki, A.H., Carniel, E., Mostafavi, E., *Plague in Iran: its history and current status*, Epidemiology and Health, 38, 2016

Shearman, J.F., *The Celto-Britons of Armorica-Maximus Clemens, Emperor of Britain, Gaul and Spain; The British Legions; Conan of Merladog, His Descendants Kings of Britagne: Continued Intercourse with the Mother – Country*, Armorican Saints and Ecclesiastics; *Their Connexions with Great Britain and Ireland in the Fifth and Sixth* Centuries, The Journal of the Royal Historical and Archaeological Association of Ireland, 5, 49, 597-630, 1882

Sharp, P.M., Hahn, B.H., *The evolution of HIV-1 and the origin of AIDS*, Phil. Trans. R. Soc, 356, 1552, 2487-2494. 2010

Shild, L., *"Go eat a bat, Chang!": An Early Look on the Emergence of Sinophobic Behavior on Web Communities in the Face of COVID-19*, arXiv preprint arXiv, 2004, 04046, 2020

Shukla, J., et al, *Amazon Deforestation and Climate Change*, Science, 247, 4948, 1322-1325, 1990

Sikora, A., *European Green Deal - legal and financial challenges of the climate change*, ERA Forum, 21, 681-697, 2021

Sivan, H., *On Foederati, Hospitalitas, and the Settlement of the Goths in A.D. 418*, The American Journal of Philology, 108, 4, 759-772, The Johns Hopkins University Press, 1987

Sołtysiak, A., *The plague pandemic and the Slavic expansion in the 6th-8th centuries*, Archaeologia Polona, Institute of Archaeology and Ethnology Polish Academy of Sciences, 44, 339-364, 2006

Southern, P., *The Roman Empire: from Severus to Constantine*, Routledge, Londra & New York, 2001

Spinoni, J., et al., *Will drought events become more frequent and severe in Europe?* Int. J. Climatol., 38, 1718-1736, 2018

Statista, *Incidence of coronavirus (COVID-19) cases in Europe as of May 9, 2021, by country*, 9th of May 2021

Stothers, R.B., *Mystery cloud of A.D. 536*, Nature, 306, 344-345, 1984

Sturluson S. *The Younger Edda; Also called Snorre's Edda, or The Prose Edda*, Perlego, ebook, 2006

Swain, A., *Challenges for water sharing in the Nile basin: changing geo-politics and changing climate*. Hydrol. Sci. J. 56, 4, 687-702, 2011

Sussman G., *Scientists Doing History: Central Africa and the Origins of the First Plague Pandemic*, Journal of World History, 26, 2, 325-354, 2015

Sukhdev, P., et al., *Green economy, developing countries success stories*, UNEP, 2010

Tacito, *La Germania*, I classici del pensiero libero Greci e Latini, BUR, Milano, 2012

Ţepelea, M., *Migration and Integration of Barbarians into the Roman Empire*, In Vertreibung, Flucht, Migration, Switzerland, Bern 2016

The Letters of Cassiodorus, translation of Thomas Hogdin, printed by Horace Hart, printer to the University, Oxford, 1886

Tovar, A., *Un obispo con nombre británico y los orígenes de la diócesis de Mondoñedo*, Habis, 3, 155-158, 1972

Triandafyllidou, A., *Migration in Greece Recent Developments in 2014*, Hellenic foundation for European and Foreign Policy, Report prepared for the OECD Network of International Migration Experts Paris, 6-8 October 2014, European University Institute, Robert Schuman Centre for Advanced Studies, 2014

Trakilović, M., *"On this path to Europe" - the symbolic role of the 'Balkan corridor' in the European migration debate* in Cultures, Citizenship and Human Rights, edited by Rosemarie Buikema, Antoine Buyse and Antonius C.G.M. Robben, Routledge, London, 2019

Tsiamis, C., Poulakou-Rebelakou, E., Androutsos, G., *The role of the Egyptian Sea and Land routes in the Justinian plague: the case of Pelusium,* Medicine and healing in the ancient Mediterranean world, edited by Demetrios Michedelis, Oxbow Books, Oxford, 2014

Tsiamis, C., Poulakou-Rebelakou, E., Petridou, E., *The Red Sea and the Port of Clysma. A Possible Gate of Justinian's Plague,* Gesnerus 66,2, 209-217, 2009

Tvauri, A., *The impact of the climate catastrophe of 536/7 AD in Estonia and neihbouring areas,* Estonian Journal of Archaeology, 18, 1, 30-56, 2014

UNHCR, IOM, *A long and winding road,* Background Paper, Regional Conference on Refugee Protection and International Migration: Mixed Movements and Irregular Migration from the East and Horn of Africa and Great Lakes Region to Southern Africa, Dar Es Salaam, 2010

UNHCR, *UNHCR Resettlement Handboo and country chapters, Refugee status and resettlement,* 2018

UNHCR, *Routes towards the Western and Central Mediterranean Sea,* January 2021

UNHCR, *Text of the 1951 Convention Relating to the Status of Refugees,* 1951

UNHCR, *The sea route to Europe: Mediterranean passage in the age of refugees,* 2015

United Nations Environment Programme, *Livelihood security: climate change, migration and conflict in the Sahel,* 2011

Vachudova, M., *Ethnopopulism and democratic backsliding in Central Europe,* East European Politics, 36, 3, 318-340, 2020

Vadas, A., Rácz, L., *Climatic Changes in the Carpathian Basin during the Middle Ages. The State of Research,* Global Environment 12, 198-227, 2013

van Baalen, S., Mobjörk, M., *A coming anarchy? Pathways from climate change to violent conflict in East Africa,* Stockholm International Peace Research Institute & The Swedish Institute of International Affairs, 2016

van Doremalen, N., et al., *Aerosol and Surface Stability of SARS-CoV-2 as Compared with SARS-CoV-1,* The New England journal of medicine, 382,16,1564-1567, 2020

Van Lancker, W., Parolin, Z., *COVID-19, school closures, and child poverty: a social crisis in the making,* The Lancet, 5, 5, 243-244, 2020

Vidal, J., *Destroyed Habitat Creates the Perfect Conditions for Coronavirus to Emerge,* COVID-19 may be just the beginning of mass pandemics, Scientific American, 18th of March 2020

Visco, G., Comandini U., Franceschi, S., *AIDS,* Universo del Corpo, Enciclopedia Treccani online, 1999

Vogt, J., *The decline of Rome,* Richard Thornton Books, London, 1967

Volkmann, A., *Climate change, environment and migration: a GIS-based study of the Roman Iron Age to the Early Middle Ages in the river Oder region,* European Journal of Post-classical Archaeologies, 5, 2015

Walter, D., et al., *Russian Twitter Accounts and the Partisan Polarization of Vaccine Discourse,* 2015-2017, AJPH open-themed research, 110, 5, 718-724, 2020

Ward, M., *Statistics on UK trade with China,* House of Commons Library, 7379, 2020

Werz, M., Hoffman, M., *Climate Change, Migration, and Conflict,* in *The Arab Spring and Climate Change,* A Climate and Security Correlations Series, Edited by Caitlin E. Werrell and Francesco Femia, Center for American Progress, Stimson, The center for Climate and Security, 2013

Widgren, M., Climate and Causation in the Swedish Iron Age: Learning from the Present to Understand the Past, Geografisk Tidsskrift – Danish Journal of Geography, 126-134, 2012

White, L.A., Mordechai, L., *Modeling the Justinianic Plague: Comparing hypothesized transmission routes,* PLoS ONE 15,4, 2020

World Metereological Organizatation, *The State of the Global Climate 2020,* 1264, 2021

Wood, I., *Before and after the Migration to Britain in The Anglo-Saxons From the Migration Period to the Eight Century,* an ethnographic perspective, edited by John Hines, Boydell Press, Woodbridge, 1997

Woodbridge-Fisher, J., et. Al., *Vegetation and land-use change in northern Europe during late antiquity: a Regional-scale pollen-based reconstruction,* Late Antique Archaeology, 11, 1, 2015

Wolfram, H., *The Roman Empire and Its Germanic Peoples,* University of California Press, Berkley, 1997

Yamanouchi, T., Takata K., Rapid change of the Arctic climate system and its global influences -

Overview of GRENE Arctic climate change research project (2011-2016), Polar Science, 2020

Yates, D., *Paradoxes of Predation in Francophone Africa,* International Journal of Political Economy, 47, 2, 2018

Yousef, K., *The vicious circle of irregular migration from Pakistan to Greece and back to Pakistan,* Hellenic Foundation for European and foreign policy, 2013

Zhang, D., *The causality analysis of climate change and large-scale human crisis*, Proceedings of the National Academy of Sciences - PNAS 108, 42, 17296-17301, 2011

Zhang, X., et al., *Trends in Middle East climate extreme indices from 1950 to 2003,* Journal of Geophysical Research, 110, 22, 2005

Zhou, P., *A pneumonia outbreak associated with a new coronavirus of probable bat origin,* Nature, 579, 7798, 270-273, 2020

Sitography

ANSA, *Migrants: Lamorgese calls to strengthen Libya border control*, 12th of May 2021
https://www.ansa.it/english/news/general_news/2021/05/12/migrants-lamorgese-calls-to-strengthen-libya-border-control_ce61c939-47a0-4750-9e35-b69992fbe9ca.html

ANSA, *Migranti: Johansson in Turchia per discutere accordo con Ue*, 6th of May 2021
https://www.ansa.it/sito/notizie/mondo/europa/2021/05/06/migranti-johansson-in-turchia-per-discutere-accordo-con-ue_9bc1b4e5-ed9d-4a9f-9e4d-349fc64dd5be.html

Al Javed, H., *Bangladesh's fatal obsession with the European dream*, Dhaka Tribune, 11th of February 2020
https://www.dhakatribune.com/bangladesh/2020/02/11/bangladesh-s-fatal-obsession-with-the-european-dream

Anderton, R. et al., *The impact of the COVID-19 pandemic on the euro area labour market*, ECB Economic Bulletin, 8, 2020
https://www.ecb.europa.eu/pub/economic-bulletin/html/eb202008.en.html

Bassetti, F., *Environmental Migrants: Up to 1 Billion by 2050*, Foresights – The CMCC observatory on climate policies and futures, 22nd of May 2019
https://www.climateforesight.eu/migrations-inequalities/environmental-migrants-up-to-1-billion-by-2050/

Basso, F., *Pacchetto clima, l'UE si divide su emissioni e costi per gli Stati*, Corriere della Sera, 26th of May 2021
https://www.corriere.it/economia/consumi/21_maggio_26/pacchetto-clima-l-ue-si-divide-emissioni-costi-gli-stati-539b6e20-bde1-11eb-a5e7-170774e96424.shtml

BBC, *Rivers and flooding: Case study: Bangladesh*
https://www.bbc.co.uk/bitesize/guides/zgycwmn/revision/4

BBC, *India Covid: Delhi builds makeshift funeral pyres as deaths climb*, 27th of April 2021
https://www.bbc.com/news/world-asia-india-56897970

Bundeszentral fur politische Bildung, *International Migration from Bangladesh*, 30th of November 2015
https://www.bpb.de/gesellschaft/migration/laenderprofile/216104/international-migration-from-bangladesh

Castellano, V., *La mappa dei porti dove sbarcano i migranti*, AGI, 01 July of 2017
https://www.agi.it/cronaca/mappa_sbarchi_migranti_porti-1918365/news/2017-07-01/

Centres for Disease Control and Prevention, *Origin of 2009 H1N1 Flu (Swine Flu): Questions and Answers*, 2009
https://www.cdc.gov/h1n1flu/information_h1n1_virus_qa.htm

Center on Budget and policy priorities, *Tracking the COVID-19 Recession's Effects on Food, Housing, and Employment Hardships*, 7th of May 2021
https://www.cbpp.org/research/poverty-and-inequality/tracking-the-covid-19-recessions-effects-on-food-housing-and

Center For Health Security, *Event 201*
https://www.centerforhealthsecurity.org/event201/

Chambers, Richard L., et al., *"Iraq"*. Encyclopedia Britannica, 11th of January 2021
https://www.britannica.com/place/Iraq

Connolly, K., *How Russia's Covid vaccine is dividing Europe*, BBC, 17th of April 2021
https://www.bbc.com/news/world-europe-56735931

Davis, S., *Europe's autumn 'the hottest on record amid world's warmest November'*, Euronews, 07th December 2020 https://www.euronews.com/2020/12/07/europe-s-autumn-the-hottest-on-record-amid-world-s-warmest-november

Dean, A., *Deforestation and climate change,* Climate Council, 21st of August 2019
https://www.climatecouncil.org.au/deforestation/

Dhaka Tribune, *One-third of Bangladesh under water in worst floods in a decade*, 15th of July 2020
https://www.dhakatribune.com/bangladesh/2020/07/15/one-third-of-bangladesh-underwater-in-worst-flood-in-decade

Ebbs, S., *Experts call for climate change debate to move beyond settled science and focus on action,* ABC News, 22nd of October 2020
https://abcnews.go.com/Politics/experts-call-climate-change-debate-move-settled-science/story?id=73739694

Emmott, R., *Russia, China sow disinformation to undermine trust in Western vaccines*: EU, Reuters, 29th of April 2021
https://www.reuters.com/world/china/russia-china-sow-disinformation-undermine-trust-western-vaccines-eu-report-says-2021-04-28/

Encyclopaedia Britannica online, "Brittany",
https://www.britannica.com/place/Brittany-region-France

Encyclopedia Britannica online, *"Cimbri"*
https://www.britannica.com/topic/Cimbri

Encyclopaedia Britannica online, *Himiar*
https://www.britannica.com/topic/Himyar

Encyclopaedia Britannica online, *Hospital*
https://www.britannica.com/science/hospital

Enciclopedia Treccani online, *Gildas il Saggio*
https://www.treccani.it/enciclopedia/gildas-il-saggio-santo/#:~:text=570%20circa).,Houat%2C%20presso%20la%20costa%20bretone.

Encyclopaedia Britannica online *Influenza, A H1N1*
https://www.britannica.com/science/influenza-A-H1N1

Enciclopedia Britannica online, Migration Period
https://www.britannica.com/event/Dark-Ages

European Centre for Disease Prevention and Control*, MERS-CoV worldwide overview*
https://www.ecdc.europa.eu/en/middle-east-respiratory-syndrome-coronavirus-mers-cov-situation-update

European Centre for Disease Prevention and Control*, Cumulative number of coronavirus (COVID-19) deaths in Europe,* Statista
https://www.statista.com/statistics/1102288/coronavirus-deaths-development-europe/

European Centre for Disease Prevention and Control, *Number of deaths among people diagnosed with HIV and/or AIDS in Europe from 2009-2019*, Statista
https://www.statista.com/statistics/654984/deaths-among-people-diagnosed-with-hiv-and-or-aids-in-europe/

European Centre for Disease Prevention and Control, *Timeline of ECDC's response to COVID-19*
https://www.ecdc.europa.eu/en/covid-19/timeline-ecdc-response

European Commission, Communication from The Commission to The European Parliament, The Council and the European Economic and Social Committee, Trade, growth and development, 27th January 2012
https://www.eea.europa.eu/policy-documents/communication-from-the-commission-to-1

European Commission, *Climate change consequences*
https://ec.europa.eu/clima/change/consequences_en#:~:text=Consequences%20for%20Europe,winter%20floods%20could%20become%20common.

European Commission, *Communication from the Commission to the European Parliament, the European Council, The Council, the European Economic and Social Committee and the Committee of the Regions - The European Green Deal,* 11th of December 2019
https://eur-lex.europa.eu/legal-content/EN/TXT/?uri=COM%3A2019%3A640%3AFIN

European Commission - CORDIS, *A major boost for HIV vaccine identification and development*, EU research result
https://cordis.europa.eu/article/id/422576-a-major-boost-for-hiv-vaccine-identification-and-development/it

European Commission, *Coronavirus: Chinese aid to the EU delivered to Italy,* 6th of April 2020
https://ec.europa.eu/commission/presscorner/detail/de/ip_20_600

European Commission, *Coronavirus: EU strengthens action to tackle disinformation,* 10th of June 2020
https://ec.europa.eu/commission/presscorner/detail/en/IP_20_1006

European Commission, *Coronavirus: European Solidarity in action,* accessed on the 20th of May 2021
https://ec.europa.eu/info/live-work-travel-eu/coronavirus-response/coronavirus-european-solidarity-action_en

European Commission, *Demographic trends of workforce*
https://knowledge4policy.ec.europa.eu/foresight/topic/changing-nature-work/demographic-trends-of-workforce_en

European Commission, *EU-Africa Business Forum: over 3 000 participants focus on the transition to green and sustainable energy in Africa,* 28th of April 2021
https://ec.europa.eu/international-partnerships/news/eu-africa-business-forum-over-3-000-participants-focus-transition-green-and-sustainable-energy_en

European Commission, *EU climate action and the European Green Deal*
https://ec.europa.eu/clima/policies/eu-climate-action_en

European Commission, *EU Emissions Trading System (EU ETS)*
https://ec.europa.eu/clima/policies/ets_en

European Commission, *EU support for vaccines*
https://ec.europa.eu/info/live-work-travel-eu/coronavirus-response/public-health/eu-vaccines-strategy_en#:~:text=Global%20cooperation,-On%2018%20September&text=On%2012%20November%202020%2C%20the,low%20and%20middle%2Dincome%20countries.

European Commission, *EU4Health 2021-2027 – a vision for a healthier European Union*
https://ec.europa.eu/health/funding/eu4health_it

European Commission, *Immigrant Key Workers: Their Contribution to Europe's COVID-19 Response,* 24th of April 2021
https://knowledge4policy.ec.europa.eu/publication/immigrant-key-workers-their-contribution-europes-covid-19-response_en

European Commission, *Impact of Covid-19 crisis on retail trade,* 2021
https://ec.europa.eu/eurostat/statistics-explained/index.php?title=Impact_of_Covid-19_crisis_on_retail_trade

European Commission, *Tackling coronavirus disinformation*, accessed on the 29th of April 2021
https://ec.europa.eu/info/live-work-travel-eu/coronavirus-response/fighting-disinformation/tackling-coronavirus-disinformation_en

European Commission, *Temporary Reintroduction of Border Control,* accessed on the 20th of May 2021
https://ec.europa.eu/home-affairs/what-we-do/policies/borders-and-visas/schengen/reintroduction-border-control_en

European Commission, *€1 billion Team Europe initiative on manufacturing and access to vaccines, medicines and health technologies in Africa,* 21st of May 2021
https://ec.europa.eu/commission/presscorner/detail/en/ip_21_2594

European Environment Agency, *Economic losses from climate-related extremes in Europe,* 16th of March 2021 https://www.eea.europa.eu/data-and-maps/indicators/direct-losses-from-weather-disasters-3/assessment-2

European Environment Agency, *Heavy precipitation in Europe,* 23rd February 2021 https://www.eea.europa.eu/data-and-maps/indicators/precipitation-extremes-in-europe-3

European Institute for gender inequality, *Covid-19 derails gender equality gains*, 05 March 2021 https://eige.europa.eu/news/covid-19-derails-gender-equality-gains

European Investment Bank, *2019-2020 EIB climate survey* https://www.eib.org/en/publications/the-eib-climate-survey-2019-2020

European Parliament, *European Parliament resolution of 16 January 2020 on ongoing hearings under Article 7(1) of the TEU regarding Poland and Hungary (2020/2513(RSP)),* 16 January 2020 https://www.europarl.europa.eu/doceo/document/TA-9-2020-0014_EN.html

European Parliament, *European Parliament resolution of 17 April 2020 on EU coordinated action to combat the COVID-19 pandemic and its consequences (2020/2616(RSP)),* 17th of April 2020 https://www.europarl.europa.eu/doceo/document/TA-9-2020-0054_EN.html

European Parliament, *German ban on medical equipment exports,*16th of March 2020 https://www.europarl.europa.eu/doceo/document/E-9-2020-001633_EN.html

European Parliament, *Fit for 55 package under the European Green Deal,* Legislative Train schedule, accessed on 26th of May 2021 https://www.europarl.europa.eu/legislative-train/theme-a-european-green-deal/package-fit-for-55

European Parliament, *Legislative train 03.2021 - 8 towards A New Policy on Migration*, 2019 https://www.europarl.europa.eu/legislative-train/api/stages/report/03-2021/theme/towards-a-new-policy-on-migration/file/jd-eu-resettlement-framework

European Parliament, *Masks intended for Italy blocked by France,* 3rd of April 2020 https://www.europarl.europa.eu/doceo/document/P-9-2020-002075_EN.html

European Parliament Think Tank, *Reform of the Dublin system,* 30th of September 2020 https://www.europarl.europa.eu/thinktank/en/document.html?reference=EPRS_BRI(2016)586639

Eurostat, *Asylum applications (non-EU) in the EU Member States, 2008–2020*, 27 April 2021 https://ec.europa.eu/eurostat/statistics-explained/index.php?title=Asylum_statistics

Eurostat, *China-EU - international trade in goods statistics*, March 2021 https://ec.europa.eu/eurostat/statistics-explained/index.php?title=China-EU_-_international_trade_in_goods_statistics&oldid=520429

Eurostat, *Early leavers from education and training,* April 2020 https://ec.europa.eu/eurostat/statistics-explained/index.php?title=Early_leavers_from_education_and_training

Eurostat, *Early leavers from education and training by sex and labour status*, dataset, April 2021 https://ec.europa.eu/eurostat/statistics-explained/index.php?title=Early_leavers_from_education_and_training

Eurostat, *Excess mortality – statistics*, accessed on 28/04/2021 https://ec.europa.eu/eurostat/statistics-explained/index.php?title=Excess_mortality_-_statistics

Eurostat, *Impact of COVID-19 on international trade by Member State*, 2021 https://ec.europa.eu/eurostat/statistics-explained/index.php?title=Impact_of_COVID-19_on_international_trade_by_Member_State

Eurostat, *Over 20% of EU population at risk of poverty or social exclusion in 2019*, 16th of October 2020 https://ec.europa.eu/eurostat/web/products-eurostat-news/-/edn-20201016-2

Fischetti, M., *Climate Change Hastened Syria's Civil War,* Scientific American, 2nd of March 2015 https://www.scientificamerican.com/article/climate-change-hastened-the-syrian-war/

Fleming, S., et al., *Tensions between EU and UK inflamed over Northern Ireland,* Financial Times, 3rd of March 2021 https://www.ft.com/content/c21da914-e51d-415d-aa94-663bd23aa5ea

Fondazione ISMU, *Migranti africani fanno impresa grazie a un percorso di formazione e mentoring – Comunicato stampa 12.10.2020,* 12th of October 2020
https://www.ismu.org/ufficio-stampa/

Frontex, *Migratory Routes, Eastern Border Route*
https://frontex.europa.eu/we-know/migratory-routes/eastern-borders-route/

Frontex, *Western Balkan route*
https://frontex.europa.eu/we-know/migratory-routes/western-balkan-route/

Frontex, *Number of illegal border-crossings to the European Union (EU) over the circular route from Albania to Greece from 2009 to 2019,* 28 April 2020
https://www.statista.com/statistics/455382/circular-route-illegal-border-crossing-to-the-eu/

Fruscione, G., *Covid-19 in the Balkans: The Virus of Authoritarianism*, ISPI, 30 April 2020
https://www.ispionline.it/it/pubblicazione/covid-19-balkans-virus-authoritarianism-25925

Furceri, D. et al., *COVID-19 will raise inequality if past pandemics are a guide*, VoxEu, 08 May 2020
https://voxeu.org/article/covid-19-will-raise-inequality-if-past-pandemics-are-guide

Garret, A., *The Refugee Crisis, Brexit, and the Reframing of Immigration in Britain,* Europe Now, Council for European Studies, 1st of August 2019
https://www.europenowjournal.org/2019/09/09/the-refugee-crisis-brexit-and-the-reframing-of-immigration-in-britain/

Gibbs, D., et al., *By the Numbers: The Value of Tropical Forests in the Climate Change Equation,* World Resource Institute, 4th October 2018
https://www.wri.org/insights/numbers-value-tropical-forests-climate-change-equation

Goldenberg, M., *Drought in the Horn of Africa is threatening 15 Million People,* UN Dispatch, 5th of August 2019
https://www.undispatch.com/drought-in-the-horn-of-africa-is-threatening-15-million-people/

Gregory, Bishop of Tours, *History of the Franks*, translated by Ernest Brehaut, Fordham University online
https://sourcebooks.fordham.edu/basis/gregory-hist.asp

Hammond, Mason. "*Trajan*". Encyclopedia Britannica, 15 Dec. 2020
https://www.britannica.com/biography/Trajan

Hedgecoe, G., Migrants reach Spain's Ceuta enclave in record numbers, BBC, 18th of May 2021
https://www.bbc.com/news/world-europe-57150051

Herring, D., *Isn't there a lot of disagreement among climate scientists about global warming?* NOAA, Climate.gov, 3rd of February 2020
https://www.climate.gov/news-features/climate-qa/isnt-there-lot-disagreement-among-climate-scientists-about-global-warming

IFAD, *Investment in Africa's rural areas will curb economic migration, IFAD President tells ministers,* 04th of July 2016
https://www.ifad.org/es/web/latest/-/news/investment-in-africa-s-rural-areas-will-curb-economic-migration-ifad-president-tells-ministers

Il Post, *I primi sbarchi di massa degli albanesi, trent'anni fa,* 7th of March 2021
https://www.ilpost.it/2021/03/07/sbarco-albanesi-brindisi-marzo-1991/

Il Post, *La crisi dei microchip è sempre più grave,* 5th of April 2021
https://www.ilpost.it/2021/04/05/carenza-microchip/

Il Post, *Le bare dei morti di COVID-19 di Bergamo portate sui camion dell'esercito,* 19th of March 2020
https://www.ilpost.it/2020/03/19/coronavirus-bare-bergamo-esercito/

ILO, *Addressing the Root causes of Migration in Ethiopia*
https://www.ilo.org/africa/technical-cooperation/WCMS_554065/lang--en/index.htm#:~:text=The%20project%20aims%20at%20supporting,migrants%20especially%20women%20and%20youth.

IMF, Regional Economic Outlook: Europe, European Commission, *Real gross domestic product growth rate forecasts in selected European countries from 2020 to 2021*, Statista, retrieved on April 28, 2021
https://www.statista.com/statistics/1102546/coronavirus-european-gdp-growth/

IOM, *Bangladesh, survey on drivers of migration and migrants' profile,* 2020
https://www.iom.int/news/iom-releases-results-largest-survey-drivers-migration-bangladesh

IOM, *Climate Change Drives Displacement of Thousands of Vulnerable People in Somalia*, 4th of January 2021
https://medium.com/@UNmigration/climate-change-drives-displacement-of-thousands-of-vulnerable-people-in-somalia-166171bb52c3

IOM, *Iraq, Migration drivers and reasons for migrating to Europe,* 2018
https://www.iom.int/news/iraqi-migration-europe-iom-report

IOM, *IOM Releases Results of Largest Survey on Drivers of Migration in Bangladesh,* 14th of July 2020
https://www.iom.int/news/iom-releases-results-largest-survey-drivers-migration-bangladesh

ISPI, *Sui migranti l'Europa non c'è,* 12th of May 2021
https://www.ispionline.it/it/pubblicazione/sui-migranti-leuropa-non-ce-30413

Istituto superiore di sanità, *Influenza pandemica A(H1N1)pdm09*, 2015
https://www.epicentro.iss.it/influenza/PandemiaH1N1

Istituto superiore di sanità, *Zika virus,* EpiCentro
https://www.epicentro.iss.it/zika/

Kington, T., *70,000 migrants waiting to cross from Libya, Italy warns,* The Times, 12th of May 2021
https://www.thetimes.co.uk/article/70-000-migrants-waiting-to-cross-from-libya-italy-warns-fspps92gb

La Repubblica, *Sant'Egidio: "Alla mensa +30% di utenti, raddoppiati i pasti da asporto. Arrivano anche nuovi poveri,* 29th of May 2020
https://roma.repubblica.it/cronaca/2020/05/29/news/coronavirus_roma_l_allarme_di_sant_egidio_alla_mensa_30_di_utenti_raddoppiati_i_pasti_da_asporto_arrivano_anche_nuovi-257934488/

LCLUC, Dryland Degradation in the Middle East
https://www.lcluc.umd.edu/hotspot/dryland-degradation-middle-east

Litzkow, J., *West Africa: How the Pandemic Reshapes Migration*, ISPI online, 17th of September 2020
https://www.ispionline.it/en/pubblicazione/west-africa-how-pandemic-reshapes-migration-27430

Lowen, M., Lampedusa: Italy's gateway to Europe struggles with migrant influx, BBC, 12th of May 2021
https://www.bbc.com/news/world-europe-57087818

Mallapaty, S., *China set to clamp down permanently on wildlife trade in wake of coronavirus,* Nature, 21st of February 2020
https://www.nature.com/articles/d41586-020-00499-2

Martuscelli, C., Gehrke, L., *Russia's Sputnik vaccine injects divisions into EU,* POLITICO, 14th of April 2021
https://www.politico.eu/article/russia-sputnik-coronavirus-vaccine-divisions-eu/

Michelot, M., *The "article 7" proceedings against Poland and Hungary: what concrete effects?* Institut Delors, 2019
https://institutdelors.eu/en/publications/__trashed/

Encyclopedia Britannica online, *Migration period*
https://www.britannica.com/event/Dark-Ages

Ministero dello Sviluppo Economico, *Sblocco delle esportazioni di dispositivi sanitari da Germania e Francia,* 15th of March 2020
https://www.mise.gov.it/index.php/it/198-notizie-stampa/2040880-sblocco-delle-esportazioni-di-dispositivi-sanitari-da-germania-e-francia

Mira, A., *Pandemia sociale. Effetto Covid e nuove emergenze: a Roma la povertà mangia le famiglie*, Avvenire, 23 July 2020
https://www.avvenire.it/attualita/pagine/effetto-covid-poveri-a-roma

Muggah, R., *In West Africa, Climate Change Equals Conflict,* Foreign Policy, 21st of February 2021
https://foreignpolicy.com/2021/02/18/west-africa-sahel-climate-change-global-warming-conflict-food-agriculture-fish-livestock/

Muller, R., *New Slovak PM seeks to mollify Russia in vaccine row*, Reuters, 9 April 2021 https://www.reuters.com/business/healthcare-pharmaceuticals/slovak-pm-we-need-secure-sputnik-v-vaccination-plan-2021-04-09/

National Low Income Housing Coalition, *30-40 Million People in America Could Be Evicted from Their Homes by the End of 2020,* 7th of August 2020
https://nlihc.org/news/30-40-million-people-america-could-be-evicted-their-homes-end-2020

NASA, *Climate Change: How Do We Know?*
https://climate.nasa.gov/

NASA, *Intense flooding in Bangladesh*
https://earthobservatory.nasa.gov/images/147057/intense-flooding-in-bangladesh#:~:text=One%20of%20the%20countries%20most,died%20as%20of%20July%2028.

NASA, *Scientific consensus: Earth climate is warming*
https://climate.nasa.gov/scientific-consensus/

NASA, *The causes of Climate Change*
https://climate.nasa.gov/causes/

Nicastro, A., *Ghali, il leader del Sahara diventa un caso diplomatico tra Marocco e Spagna,* Corriere della Sera, 19th of May 2021
https://www.corriere.it/esteri/21_maggio_19/ghali-leader-sahara-diventa-caso-diplomatico-marocco-spagna-89975b50-b883-11eb-86a2-256e95d23aef.shtml

Nunez, C., *Deforestation explained,* National Geographic, 7th of February 2019
https://www.nationalgeographic.com/environment/article/deforestation

OCHA, *Daily Noon Briefing Highlights: Bangladesh*, 27th of July 2020
https://www.unocha.org/story/daily-noon-briefing-highlights-bangladesh-0

OECD, *The territorial impact of COVID-19: Managing the crisis across levels of government,* 2020
https://www.oecd.org/coronavirus/policy-responses/the-territorial-impact-of-covid-19-managing-the-crisis-across-levels-of-government-d3e314e1/

O'Neill, A., *Estimated death toll of the Plague of Justinian 541-767*, Statista LUISS, April 2020
https://www.statista.com/statistics/1114242/justinian-plague-estimates/

Palm, A., *The Italy-Libya Memorandum of Understanding: The baseline of a policy approach aimed at closing all doors to Europe?* EU Immigration and Asylum Law and Policy Droit et Politique de l'Immigration et de l'Asile de l'UE, 02 October 2017
https://eumigrationlawblog.eu/the-italy-libya-memorandum-of-understanding-the-baseline-of-a-policy-approach-aimed-at-closing-all-doors-to-europe/

Putoto, G., *Environmental Vulnerability: South Sudan's Endgame*, ISPI online, 23rd of March 2018
https://www.ispionline.it/it/pubblicazione/environmental-vulnerability-south-sudans-endgame-19956

Qayum, N., *Chasing the Dubai Dream in Italy: Bangladeshi Migration to Europe*, Migration Policy Institute, 5th of October 2017
https://www.migrationpolicy.org/article/chasing-dubai-dream-italy-bangladeshi-migration-europe

Treccani.it – Vocabolario Treccani on line, Istituto dell'Enciclopedia Italiana, *Quarantena*
https://www.treccani.it/vocabolario/quarantena#:~:text=Periodo%20di%20quaranta%20giorni%3A%20indulgenza,germi%20di%20malattie%20infettive%2C%20spec.

Reuters, *Italia, 2020 ha registrato mortalità più alta da Dopoguerra a causa del Covid – Istat,* 5 of March 2021
https://www.reuters.com/article/salute-coronavirus-italia-morti-idITKCN2AX21S

Roser, M., Ritchie, H., *HIV / AIDS*, Published online at OurWorldInData.org. Retrieved from: https://ourworldindata.org/hiv-aids, 2018
https://ourworldindata.org/hiv-aids

Sant'Egidio, *Dati e proposte per mettere fine alla "eterna zona rossa" per gli anziani ricoverati in strutture assistenziali*, 28 April 2021
https://www.santegidio.org/pageID/30284/langID/it/itemID/41933/Dati-e-proposte-per-mettere-fine-alla-eterna-zona-rossa-per-gli-anziani-ricoverati-in-strutture-assistenziali.html

Sant'Egidio, *South Sudan: together with Sant'Egidio major step forward towards Peace,* 11th of March 2021 https://www.santegidio.org/pageID/30284/langID/en/itemID/41000/South-Sudan-together-with-Sant-Egidio-major-step-forward-towards-Peace.html

Saha, S., How climate change could exacerbate conflict in the Middle East, Atlantic Council, 14th of May 2019
https://www.atlanticcouncil.org/blogs/menasource/how-climate-change-could-exacerbate-conflict-in-the-middle-east/

Sandle, P., Chalmers, J., *EU blames AstraZeneca as vaccine battle with UK deepens,* Reuters, 22nd of March 2021
https://www.reuters.com/article/us-health-cornavirus-britain-eu-idUKKBN2BE1BQ

Simmons, I.G., *Margins of the East Fen: Historic Landscape Evolution*, Durham University, Appendix 2.1.A Weather, Climate, Sea level and floods in Medieval Times
https://www.dur.ac.uk/east-lincs-history/investigations/the-saxon-shore-the-vikings-and-domesday-book/appendix-2.1.a/

Slattery, G., *Aspirin in honey: dubious COVID-19 'cures' spread in Brazil*, Reuters, 17th of June 2020 https://www.reuters.com/article/us-health-coronavirus-brazil-fakenews-idUSKBN23O2VA

Sobotka, B., *Two-speed Europe is a mistake,* POLITICO, 6th of December 2017 https://www.politico.eu/article/bohuslav-sobotka-oped-two-speed-europe-is-a-mistake/

Specia, M., *383,000: Estimated Death Toll in South Sudan's War*, The New York Times, 26th of September 2018 https://www.nytimes.com/2018/09/26/world/africa/south-sudan-civil-war-deaths.html

Terry, K., *The EU-Turkey Deal, Five Years On: A Frayed and Controversial but Enduring Blueprint,* The on-line journal of the Migration Policy Institute, 8 April 2021
https://www.migrationpolicy.org/article/eu-turkey-deal-five-years-on

The Annals of Ulster, Electronic edition compiled by Pádraig Bambury and Stephen Beechinor, CELT https://celt.ucc.ie//published/T100001A/index.html

The Arctic Council, *The impact of Covid-19 on Saami communities*, 2020 https://arctic-council.org/en/news/the-impact-of-covid-19-on-saami-communities/

The EU's Assembly of Regional and Local Representatives, *The Covid-19 crisis can deepen Europe's demographic decline, regions and cities warn,* 15th of October 2020
https://cor.europa.eu/en/news/Pages/eu-demographic.aspx

The World Bank, *COVID-19 to Add as Many as 150 Million Extreme Poor by 2021,* 7th of October 2020 https://www.worldbank.org/en/news/press-release/2020/10/07/covid-19-to-add-as-many-as-150-million-extreme-poor-by-2021

The World Bank, *COVID-19 to Plunge Global Economy into Worst Recession since World War II,* 8th of June 2020
https://www.worldbank.org/en/news/press-release/2020/06/08/covid-19-to-plunge-global-economy-into-worst-recession-since-world-war-ii#:~:text=WASHINGTON%2C%20June%208%2C%202020%20%E2%80%94,shrink%20by%205.2%25%20this%20year.

Thomas, P., et al., *Disaster in motion: 3.4 million travelers poured into US as coronavirus pandemic erupted*, ABC News, 2nd of April 2020
https://abcnews.go.com/Politics/disaster-motion-34-million-travelers-poured-us-coronavirus/story?id=69933625

Triggle, N., *Covid: 2020 saw most excess deaths since World War Two*, BBC, 12 of January 2021 https://www.bbc.com/news/uk-55631693

UNAIDS, *Global HIV & AIDS statistics — 2020 fact sheet*
https://www.unaids.org/en/resources/fact-sheet

UNAIDS, *Who we are*
https://www.unaids.org/en/whatwedo

UNAIDS, *Global HIV & AIDS statistics — 2020 fact sheet*
https://www.unaids.org/en/resources/fact-sheet

UNCTAD, *Green energy could help developing countries diversify their economies,* 12th of October 2017 https://unctad.org/news/green-energy-could-help-developing-countries-diversify-their-economies

UNCTAD, *Promoting refugee and migrant entrepreneurship in East Africa, the Andean region, and the Middle East*
https://unctad.org/project/promoting-refugee-and-migrant-entrepreneurship-east-africa-andean-region-and-middle-east

UNDESA, *Total population across the Middle East and North Africa from 1990 to 2050 (in million)* [Graph]. In Statista
https://www.statista.com/statistics/978535/mena-total-population/

University of Oxford, *Poverty and inequality surge across Europe in the wake of COVID-19*, 29 October 2020
https://www.ox.ac.uk/news/2020-10-29-poverty-and-inequality-surge-across-europe-wake-covid-19

University of Notre Dame, *ND-GAIN,* Country Index
https://gain.nd.edu/our-work/country-index/rankings/

UNHCR, *Desperate journeys, Refugees and migrants entering and crossing Europe via the Mediterranean and Western Balkans routes,* 2017
https://www.unhcr.org/news/updates/2017/2/58b449f54/desperate-journeys-refugees-migrants-entering-crossing-europe-via-mediterranean.html

UNHCR, *Figures at a Glance,* 18th June 2020
https://www.unhcr.org/ph/figures-at-a-glance#:~:text=UNHCR's%20annual%20Global%20Trends%20Report,countries%20that%20are%20most%20affected.

UNHCR, Operations, East and Horn of Africa and Great Lakes, *Somalia*
https://data2.unhcr.org/en/situations/horn

UNHCR, *Operational portal, Refugee Situation, DRC Situation*
https://data2.unhcr.org/en/situations/drc

UNHCR, Operational portal, Refugee Situation, *Ethiopia*
https://data2.unhcr.org/en/country/eth

UNHCR, Operational portal, Refugee Situation, *Horn of Africa Somalia situation*
https://data2.unhcr.org/en/situations/horn

UNHCR, *Operation Portal, Refugee situations, Mediterranean Situation, Greece*
https://data2.unhcr.org/en/situations/mediterranean

UNHCR, *Operation Portal, Refugee situations, Mediterranean Situation, Italy*
https://data2.unhcr.org/en/situations/mediterranean

UNHCR, Operational Portal, Refugee Situation, Mediterranean Situation, *Spain*
https://data2.unhcr.org/en/situations/mediterranean

UNHCR, *Refugee Data Finder*
https://www.unhcr.org/refugee-statistics/

UNHCR, *UNHCR airlift supplies to assist thousands of Somalis displaced by Cyclone Gati,* 18th December 2020
https://www.unhcr.org/news/briefing/2020/12/5fdc6cd24/unhcr-airlift-supplies-assist-thousands-somalis-displaced-cyclone-gati.html

Union of concerned scientists, *Each Country's Share of CO2 Emissions*, accessed on 21st of May 2021
https://www.ucsusa.org/resources/each-countrys-share-co2-emissions

United Nations Climate change, *"Climate Commitments Not on Track to Meet Paris Agreement Goals" as NDC Synthesis Report is Published,* 26th of February 2021
https://unfccc.int/news/climate-commitments-not-on-track-to-meet-paris-agreement-goals-as-ndc-synthesis-report-is-published

United Nations Climate Change, *Climate Change Is an Increasing Threat to Africa,* 27th of October 2020
https://unfccc.int/news/climate-change-is-an-increasing-threat-to-africa#:~:text=UN%20Climate%20Change%20News%2C%2027,devoted%20exclusively%20to%20the%20continent.

Villa, M., Fact Checking: migrazioni (e Covid-19), ISPI online, 29th of July 2020
https://www.ispionline.it/it/pubblicazione/fact-checking-migrazioni-e-covid-19-27058

Vivekananda, J., *Climate Change and the Conflict Trap in Lake Chad,* ISPI online, 19th of March 2020
https://www.ispionline.it/it/pubblicazione/climate-change-and-conflict-trap-lake-chad-25442

Wagstyl, S., *Climate change is becoming less a battle of nations than rich vs poor,* Financial Times, 21st of May 2021
https://www.ft.com/content/4788beae-9035-4449-b5cd-200dc7b6ea9d

WHO-convened Global Study of Origins of SARS-CoV-2: China Part, Joint Report, 2021
https://www.who.int/publications-detail-redirect/who-convened-global-study-of-origins-of-sars-cov-2-china-part

WHO, *WHO Coronavirus (COVID-19) Dashboard*
https://covid19.who.int/

WHO, *Ebola virus disease*
https://www.who.int/news-room/fact-sheets/detail/ebola-virus-disease

WHO, *Ebolavirus disease,* Fact sheet n° 103, 2014

WHO, *Ebola outbreak 2021-* N'Zerekore, Guinea
https://www.who.int/emergencies/situations/ebola-2021-nzerekore-guinea

WHO, *Ebola outbreak in West Africa and the risk to Europe*, WHO Regional Office for Europe
https://www.euro.who.int/en/health-topics/health-emergencies/ebola-outbreak-2014/ebola-outbreak-in-west-africa-and-the-risk-to-europe#:~:text=Ebola%20outbreak%20in%20West%20Africa%20and%20the%20risk%20to%20Europe,-AddThis%20Sharing%20Buttons&text=The%20outbreak%20started%20in%20Guinea,other%20known%20Ebola%20outbreaks%20combined.

WHO, *Middle East respiratory syndrome coronavirus (MERS-CoV)*
https://www.who.int/health-topics/middle-east-respiratory-syndrome-coronavirus-mers#tab=tab_1

WHO, *Middle East respiratory syndrome coronavirus (MERS-CoV) and the risk to Europe*, WHO Regional Office for Europe
https://www.euro.who.int/en/health-topics/health-emergencies/middle-east-respiratory-syndrome-coronavirus-mers-cov-and-the-risk-to-europe

WHO, *Situation Updates – SARS*, Emergencies preparedness, response
https://www.who.int/csr/sars/archive/en/

WHO, *Summary of probable SARS cases with onset of illness from 1 November 2002 to 31 July 2003*
https://www.who.int/publications/m/item/summary-of-probable-sars-cases-with-onset-of-illness-from-1-november-2002-to-31-july-2003

WHO, *Pandemic (H1N1) 2009 - update 112,* 2010
https://www.who.int/csr/don/2010_08_06/en/

WHO, *Timeline: WHO's COVID-19 response*
https://www.who.int/emergencies/diseases/novel-coronavirus-2019/interactive-timeline

WHO, *Zika virus in the Western Pacific*
https://www.who.int/westernpacific/health-topics/zika#:~:text=Zika%20virus%20in%20the%20Western%20Pacific&text=These%20symptoms%20are%20usually%20mild,trigger%20of%20Guillain%2DBarr%C3%A9%20syndrome.

WHO, *Zika virus outbreak and the risk to Europe,* WHO Regional Office for Europe, 2016
https://www.euro.who.int/en/health-topics/health-emergencies/zika-virus/zika-virus-outbreak-and-the-risk-to-europe2

Yacoubian, M., *What is Russia's Endgame in Syria?* United States Institute of Peace, 16th of February 2021
https://www.usip.org/publications/2021/02/what-russias-endgame-syria

Zhang, S., *An Ancient Case of the Plague Could Rewrite History*, The Atlantic, 6 December 2018
https://www.theatlantic.com/science/archive/2018/12/4900-year-old-case-plague-sweden/577315/

Finito di stampare nel mese di marzo 2022
per conto de «L'ERMA» di BRETSCHNEIDER®
da Services4Media s.r.l. - Viale Caduti di Nassiriya, 39 - 70124 Bari